Stop Faking It!

Finally Understanding Science   So You Can Teach It

FORCE AND MOTION

Stop Faking It!

Finally Understanding Science So You Can Teach It

FORCE AND MOTION

NATIONAL SCIENCE TEACHERS ASSOCIATION

National Science Teachers Association
Arlington, Virginia

NATIONAL SCIENCE TEACHERS ASSOCIATION

Claire Reinburg, Director
Judy Cusick, Associate Editor
Carol Duval, Associate Editor
Betty Smith, Associate Editor

ART AND DESIGN Linda Olliver, Director
   Brian Diskin, Illustrator
NSTA WEB Tim Weber, Webmaster
PERIODICALS PUBLISHING Shelley Carey, Director
PRINTING AND PRODUCTION Catherine Lorrain-Hale, Director
   Nguyet Tran, Assistant Production Manager
   Jack Parker, Desktop Publishing Specialist
PUBLICATIONS OPERATIONS Erin Miller, Manager
*sci*LINKS Tyson Brown, Manager

NATIONAL SCIENCE TEACHERS ASSOCIATION
Gerald F. Wheeler, Executive Director
David Beacom, Publisher

Featuring sciLINKS®—a new way of connecting text and the Internet. Up-to-the-minute online content, classroom ideas, and other materials are just a click away. Go to page x to learn more about this new educational resource.

Force and Motion:Stop Faking It! *Finally Understanding Science So You Can Teach It*
   NSTA Stock Number: PB169X1
05   04   5   4   3
   ISBN: 0–87355–209–1
   Library of Congress Control Number: 2002102569
   Printed in the USA by Kirby Lithographic Company, Inc.
   Printed on recycled paper

NSTA is committed to publishing quality materials that promote the best in inquiry-based science education. However, conditions of actual use may vary and the safety procedures and practices described in this book are intended to serve only as a guide. Additional precautionary measures may be required. NSTA and the author(s) do not warrant or represent that the procedures and practices in this book meet any safety code or standard or federal, state, or local regulations. NSTA and the author(s) disclaim any liability for personal injury or damage to property arising out of or relating to the use of this book including any of the recommendations, instructions, or materials contained therein.

Contents

Preface

When I was back in college, there was a course titled Physics for Poets. At a school where I taught physics, the same kind of course was referred to by the students as Football Physics. The theory behind having courses like these was that poets and/or football players, or basically anyone who wasn't a science geek, needed some kind of watered-down course because most of the people taking the course were—and this was generally true—SCARED TO DEATH OF SCIENCE.

In many years of working in education, I have found that the vast majority of elementary school teachers, parents who home-school their kids, and parents who just want to help their kids with science homework fall into this category. There are lots of "education experts" who tell teachers they can solve this problem by just asking the right questions and having the kids investigate science ideas on their own. These experts say you don't need to understand the science concepts. In other words, they're telling you to fake it! Well, faking it doesn't work when it comes to teaching *anything*, so why should it work with science? Like it or not, you have to understand a subject before you can help kids with it. Ever tried teaching someone a foreign language without knowing the language?

The whole point of the *Stop Faking It!* series of books is to help you understand basic science concepts and to put to rest the myth that you can't understand science because it's too hard. If you haven't tried other ways of learning science concepts, such as looking through a college textbook, or subscribing to *Scientific American*, or reading the incorrect and oversimplified science in an elementary school text, please feel free to do so and then pick up this book. If you find those other methods more enjoyable, then you really are a science geek and you ought to give this book to one of us normal folks. Just a joke, okay?

Just because this book series is intended for the nonscience geek doesn't mean it's watered-down material. Everything in here is accurate, and I'll use math when it's necessary. I will stick to the basics, though. My intent is to provide a clear picture of underlying concepts without all the detail on units, calculations, and intimidating formulas. You can find that stuff just about anywhere. Also, I'll try to keep it lighthearted. Part of the problem with those textbooks (from elementary school through college) is that most of the authors and teachers who use them take themselves way too seriously. I can't tell you the number of times I've written science curriculum only to have colleagues tell me it's "too flip" or "You know, Bill, I just don't think people will get this joke." Actually, I don't really care if you get the jokes either, as long as you manage to learn some science here.

Speaking of learning the science, I have one request as you go through this book. There are two sections titled *Things to do before you read the science stuff* and *The science stuff*. The request is that you actually DO all the "things to do" when I ask you to do them. Trust me, it'll make the science easier to understand, and it's not like I'll be asking you to go out and rent a superconducting particle accelerator. Things around the house should do the trick.

As you go through this book, you'll notice that just about everything is measured in *Système Internationale* (SI) units, such as meters, kilometers, and kilograms. You might be more familiar with the term *metric units*, which is basically the same thing. There's a good reason for this—this is a science book and scientists the world over use SI units for consistency. Of course, in everyday life in the United States, people use what are commonly known as English units (pounds, feet, inches, miles, and the like). The only time I'll use English units is when it would be silly to do otherwise, such as giving the speed of a car in miles per hour rather than kilometers per hour. If SI units really baffle you, there are a few easy conversions you can make to English units. There are about 1.6 kilometers in a mile, there are about 3 feet in a meter, and a 1 kilogram mass has a weight of about 2 pounds.

The book you have in your hands, *Force and Motion*, deals with—get ready for a surprise—force and motion. I'll address how to describe the motion of things as well as how Isaac Newton viewed the relationship between forces, and between forces and changes in motion. Then, after two chapters on the force of gravity and circular motion, I'll do my best to pull most of the ideas together by describing how you would go about getting a rocket ship to the Moon. You will notice that this book is not laid out the way these topics might be addressed in a high school or college textbook. That's because this isn't a textbook. You can learn a great deal of science from this book, but it's not a traditional approach.

One more thing to keep in mind: You actually CAN understand science. It's not that hard when you take it slowly and don't try to jam too many abstract ideas down your throat. Jamming things down your throat, by the way, seemed to be the philosophy behind just about every science course I ever took. Here's hoping this series doesn't continue that tradition.

Acknowledgments

The *Stop Faking It!* series of books is produced by the NSTA Press: Claire Reinburg, director; Carol Duval, project editor; Linda Olliver, art director; Catherine Lorrain-Hale, production director. Linda Olliver designed the cover from an illustration provided by artist Brian Diskin, who also created the inside illustrations.

This book was reviewed by Lynn Cimino-Hurt (Flint Hill School, Virginia); Terri Matteson (Estrella Mountain Elementary School, Arizona); and Daryl Taylor (Williamstown High School, New Jersey). Thanks also to Beth Daniels for supporting the publication of this series.

About the Author

Bill Robertson is a science education writer, teaches online math and physics for the University of Phoenix, and reviews and edits science materials. His numerous publications cover issues ranging from conceptual understanding in physics to bringing constructivism into the classroom. Bill has developed K–12 science curricula, teacher materials, and award-winning science kits for Biological Sciences Curriculum Study, The United States Space Foundation, The Wild Goose Company, and Edmark. Bill has a master's degree in physics and a Ph.D. in science education.

About the Illustrator

The soon to be out-of-debt humorous illustrator Brian Diskin grew up outside of Chicago. He graduated from Northern Illinois University with a degree in commercial illustration, after which he taught himself cartooning. His art has appeared in many books, including *The Golfer's Personal Trainer* and *5 Lines: Limericks On Ice*. You can also find his art in newspapers, on greeting cards, on T-shirts, and on refrigerators. At any given time he can be found teaching watercolors and cartooning, and hopefully working on an ever-expanding series of *Stop Faking It!* books. You can view his work at *www.briandiskin.com*.

The *Stop Faking It* series brings you *sci*LINKS, a new project that blends the two main delivery systems for curriculum—books and telecommunications—into a dynamic new educational tool for children, their parents, and their teachers. *sci*LINKS links specific science content with instructionally-rich Internet resources. *sci*LINKS represents an enormous opportunity to create new pathways for learners, new opportunities for professional growth among teachers, and new modes of engagement for parents.

In this *sci*LINKed text, you will find an icon near several of the concepts being discussed. Under it, you will find the *sci*LINKS URL (*www.scilinks.org*) and a code. Go to the *sci*LINKS website, sign in, type the code from your text, and you will receive a list of URLs that are selected by science educators. Sites are chosen for accurate and age-appropriate content and good pedagogy. The underlying database changes constantly, eliminating dead or revised sites or simply replacing them with better selections. The *sci*LINKS search team regularly reviews the materials to which this text points, so you can always count on good content being available.

The selection process involves four review stages:

1. First, a cadre of undergraduate science education majors searches the World Wide Web for interesting science resources. The undergraduates submit about 500 sites a week for consideration.

2. Next, packets of these web pages are organized and sent to teacher-webwatchers with expertise in given fields and grade levels. The teacher-webwatchers can also submit web pages that they have found on their own. The teachers pick the jewels from this selection and correlate them to the National Science Education Standards. These pages are submitted to the *sci*LINKS database.

3. Scientists review these correlated sites for accuracy.

4. NSTA staff approve the web pages and edit the information provided for accuracy and consistent style.

*sci*LINKS is a free service for textbook and supplemental resource users, but obviously someone must pay for it. Participating publishers pay a fee to NSTA for each book that contains *sci*LINKS. The program is also supported by a grant from the National Aeronautics and Space Administration (NASA).

Newton's First One

This first chapter deals with one of the most basic principles of motion, which happens to be known as Newton's first law. Not coincidentally, it has something to do with Isaac Newton. It's a nice law to start out with because it doesn't require any math and, after all, it is the first law. Just about every textbook I've seen spends very little time on Newton's first law, presumably because it's so basic and obvious. You'll find out, though, that at least part of the law is far from obvious.

This chapter is also our first step toward understanding what science knowledge you need to plan a trip to the Moon. Not that you necessarily wanted to go to the Moon, but since each chapter builds on previous ones, the Moon-trip chapter (the last one) seemed a fun way to tie everything together. So if you thought you were going to skip around and maybe hit the Moon trip before going through everything else, forget about it!

Newton's 1st

Things to do before you read the science stuff

Center an index card over the top of a glass, and place a coin in the middle of the index card (on top might be a good place). Using just one or two fingers, flick the card from the side.

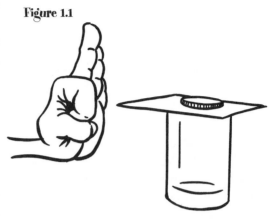

Figure 1.1

If you do your flicking just right, the card should fly to the side and the coin should fall into the glass. Ah, yes, science is magic.

Now place the coin on a flat, level surface. Watch it for a while, say about five minutes. Notice whether or not it starts moving all by itself.

The science stuff

You have just demonstrated for yourself the first part of Mr. Newton's first law, which is that **objects tend to remain at rest unless you hit them**. Okay, so Newton didn't say it that way. I'll give you the fancy language later. Hopefully the coin didn't move as you watched it on the flat surface. The index card didn't stay where it was because you hit it, but the coin on top of the card stayed where it was (because you didn't hit it), at least until gravity pulled it into the glass.

Easy stuff, right? Told you science wasn't hard.

More things to do before you read more science stuff

Grab a few things from around the house—some you can roll across the floor (a ball, a toy car, a rolling pin, a glass) and some you can slide across the floor (a book, a block of wood, that fruitcake left over from the holidays—yes, it *is* frightening that Uncle Wally hasn't left yet.) Roll and slide these things across various *level* surfaces, such as tile and carpet. Watch carefully what happens. What do they do? Speed up, slow down, stop, keep going forever? Go outside and try it where you've got lots of room to check out that keep-going-forever thing.

More science stuff

Time to introduce the second part of Newton's first law, which is that **objects in motion tend to stay in motion until something hits them** (again my words, not Isaac's). That means that once something is moving, it will keep moving forever, without anything pushing it along, until something else pushes or pulls on it. Of course you just saw that happen with all the objects you were rolling and sliding across the floor. Then again, maybe you didn't see that happen.

In discussing this idea with folks, I usually challenge them to give an example of *any* object in their daily lives that keeps on moving forever without anything pushing it along. Some clever person always says "a two year old." Nice try, but a two year old doesn't move by him or herself. The kid needs a floor doing some pushing to even walk (see Chapter 5), and this is fueled by all the sugar he or she had for breakfast. The deep thinkers in the audience usually mention "time" as something that moves forever without a push. True, but time doesn't count as an object. We're talking your garden-variety touchable things. Finally, someone always mentions that things keep cruising along without a push in outer space. It always amazes me that I run into so many people who consider a trip to outer space to be a part of their everyday life.

So if Newton's first law doesn't apply to everyday life, what good is it? Well, it really does apply to everyday life, but we'll have to explore a bit further to see how. The reason I bring up the issue at this point is to show that science concepts don't always make sense when you first come across them. That can get in the way of understanding what they're about. I guess that what goes through lots of people's heads is something like this: "Okay, this concept doesn't fit with my everyday experience, so that means there's not a lot of common sense in science. There must be some strange reasoning process going on here, and I don't get it. I suppose I can just memorize this stuff and get through the best I can." I don't want you to think that way, so instead of presenting science concepts as facts that you would understand if you only had a brain, I'll try to show you how those wacky scientists came up with the ideas in the first place. The next activity will get us moving along that line. In the meantime, keep the following in mind.

People who develop science concepts *make them up*. These ideas are *not* handed down from deities on high. Science concepts hang around, not because they're always simple and obvious, but because they work, meaning that they help us understand and predict things. That said, it's not as if you can come up with any old explanation and call it a valid scientific theory. There are conventions for evaluating theories to determine how good they are. Suffice it to say that the theories presented in this book have been around a while and have stood the test of time.

Even more things to do before you read more science stuff

Grab a ruler or yardstick, a marble or a ball bearing, and about a meter-long section of your kid's Hot Wheels track. If you don't have access to kids' toys, just use anything you can find that's flexible and will allow a marble to roll along it. What works well is a section of clear, plastic tubing (try the hardware or plumbing supply store) and a ball bearing that's small enough to roll freely inside the tubing.

Find a friend or family member to help you with this next part. Hold the track in a U shape[1] so the lowest part just touches a table top or a floor. (Check out Figure 1.2.) Now measure the *vertical* distance from the floor or table to one end of the track. For the directionally challenged, that vertical distance is shown in Figure 1.3.

Figure 1.2

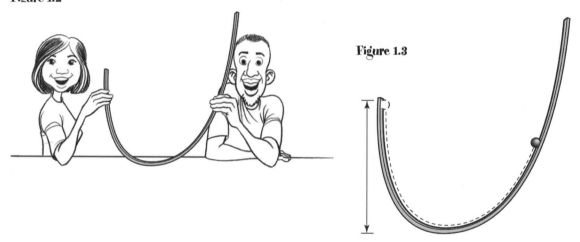

Figure 1.3

If your memory isn't great, write this distance down. You'll need to keep this one side of the track at that same vertical distance as you do the next few things. With your accomplice helping you hold the track in a U shape with the bottom of the U touching the table or floor, and holding your end at the vertical distance you've measured, drop the marble at the top of that end of the track (Figure 1.4).

---

[1] There should be quite a bit left over on one end, so I guess this is really more of a J shape.

Figure 1.4

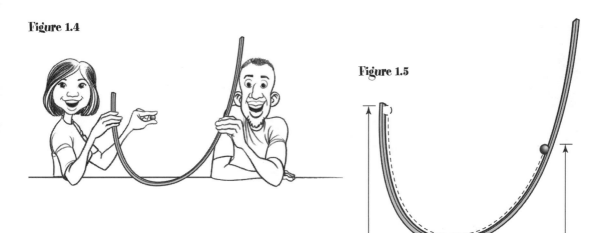

Figure 1.5

Watch what the marble does. It should roll back and forth for a while and finally come to rest. Now do this a few more times, but notice where the marble stops *the first time* before it goes back in the other direction. When you've done it enough times that the marble seems to stop in just about the same place each time, measure the vertical distance from the table or floor to this point on the track (Figure 1.5).

Now that you're good at this procedure, you get to repeat it. Only this time change the shape of the track so the second side of the U is lowered a bit. (Look at Figure 1.6.) As you repeat things, make sure your starting height is the same as it was the first time you did it (Figure 1.7).

Figure 1.6

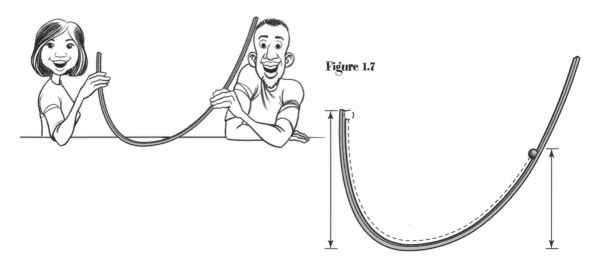

Figure 1.7

Figure 1.8

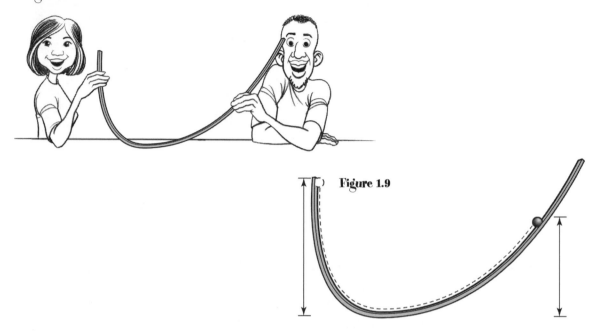

Figure 1.9

Now lower the second side of the U even more and repeat (Figures 1.8 and 1.9).

If you were doing this in a science course or were teaching a science course, the last thing anyone should do at this point is talk about what *should* have just happened in your little experiment. But hey, this is a book and I've got to get on with things, so here's what should have happened. No matter what the shape of the track, the marble should have risen to just about the *same vertical distance* each time. You probably didn't get exactly the same distance each time, but if you were within a few centimeters, that's good enough.

Now assuming you did get the result that the marble rises to the same height each time, no matter what the shape of the track, see if you can answer the following question. If the track were shaped like an L (see Figure 1.10) and the bottom part of the L were infinitely long (going on forever), how far would the marble travel before turning around and going back?

Figure 1.10

Even more science stuff

Because your experiment worked out just exactly as I told you it should (science experiments always work, don't they?), you could make up a general rule for how the marble behaved. That rule might go like this: No matter what the shape of the track, the marble will travel along the far side until it reaches a vertical distance of (*enter your distance here*), at which point it will turn around and go back. If you buy that rule, then your answer to the question about the L-shaped track has to be something like this: Since the marble will travel until it reaches that certain vertical distance, and with an L-shaped track the marble will *never* reach that vertical distance, the marble will travel on forever.

All right, so maybe you wouldn't have reached that conclusion. If it makes sense to you now, that's enough. Turns out it made sense to a guy named Galileo (the activity you just did is often referred to as Galileo's ramps) and then Newton just incorporated it into his first law. So that's how the motion part of Newton's first law came about. Later I'll get to why it doesn't seem to work so well in the real world. For now, we're ready for a complete statement of Newton's first law, which is that **things tend to keep on doing whatever they're doing until something else hits them**. This means that any time there's a change in an object's motion, a push or hit from something else caused it. Changes in motion refer to starting, stopping, speeding up, slowing down, and even just changing direction. There are some examples at the end of this chapter.

Topic: Newton's first law

Go to: *www.scilinks.org*

Code: SFF01

Topic: forces

Go to: *www.scilinks.org*

Code: SFF02

It can get to be a real pain to say things like, "Hey, Wilma, looks to me like that thing over there tends to keep on doing what it's doing more so than that other thing over there." So there's a special term that refers to an object's tendency to keep on doing what it's doing—**inertia**. It's hard to change the motion of an object with lots of inertia and it's easy to change the motion of an object with very little inertia. Objects with very little inertia: mosquitoes, black-eyed peas, and dust mites. Objects with lots of inertia: elephants, mountains, and government-run agencies.

The term "hit" is also a bit too restrictive. It turns out that you can change an object's motion by pushing it or pulling it or nudging it or blowing on it or any number of other things, all of which are referred to as **forces**.

Armed with all the "proper" terms, maybe Newton's original wording of the first law might not seem so strange: **"Every object persists in its state of rest or of uniform motion in a straight line unless it is compelled to change that state by forces impressed on it."** You go, Isaac.

Now what about the real world, where everything comes to rest? Newton's explanation is that things come to rest, not because they naturally do that, but because forces act on them to bring them to rest. A common one is the force of **friction**, which is the push things give to each other when they rub together.

Figure 1.11

In Newton's world, if you could eliminate friction, then you'd see more things keep on moving in a straight line. If you played air hockey as a kid, then you saw what can happen when there's very little friction around. One push-off doesn't help you slide across a carpet very well because there's a lot of friction between your feet and the carpet. That same shove on a pair of ice skates, though, will get you a lot farther because there's not much friction around.

Just so we're straight on things, Newton's first law *does* hold true in the real world, as long as you can account for all the forces, like friction, that might affect things. You'll probably go nuts trying to understand the first law by thinking about situations where there aren't any forces around, because there are almost always lots of forces around. If you really want to cling to the idea that objects tend to come to rest rather than stay in motion, though, you're in good company. That's what Aristotle thought, and word is he wasn't a stupid guy.

Chapter Summary

- Objects tend to keep on doing whatever they're doing (staying at rest or staying in motion) unless something else exerts a force on them. This is known as Newton's first law.

- Inertia is the name given to an object's tendency to keep doing what it's doing.

- Newton's first law only makes sense when *all* forces, including friction, are accounted for.

Applications

1. You throw a rock and it hits the ground. What does Isaac say about this? Probably nothing, because he's been dead for a really long time. But how does his first law apply?

    Well, while the rock is in your hand, its inertia just sort of keeps it there until a force acts on it. You provide that force by throwing the rock. A common mistake people make is to think that the force you exerted on the

Figure 1.12

rock must be staying with it; otherwise it wouldn't keep moving. But according to the first law, the rock will keep moving because of its inertia. It doesn't need a force to keep doing what it's already doing. Of course the rock doesn't keep doing what it's doing. It immediately starts to fall to the ground, which means a change in direction and, if you observe carefully, an increase in speed. The force that's causing this change in motion is the Earth's gravity (more on gravity later). When the rock hits the ground it stops—another change in motion. This time it's the ground exerting a force on the rock to cause the change. By the way, if the idea of an inanimate object like the ground exerting a force on something bothers you, you're going to be bothered at least until Chapter 5.

2. You and the dog are riding in the car when you slam on the brakes. Rover, who was in the back, ends up licking your face. Why? Well, you and the car and Rover are cruising along at 30 miles per hour, all of you dutifully obeying Newton's first law. When you hit the brakes, friction (a force) between the road and the tires changes the motion of the car from moving to not moving. Your seat belt exerts a force on you and changes your motion from moving to not moving. Rover doesn't have a seat belt or anything else to exert a significant force on him, so

Figure 1.13

he keeps doing what he was doing, which is traveling in a straight line at 30 miles per hour.

3. Two people push equally hard on opposite sides of a large couch and the couch doesn't go anywhere.

Figure 1.14

Because of the couch's inertia, it's going to stay at rest until a force is exerted on it, right? But aren't the two people exerting a force on it? Yes and no. Because the forces are equal and they act in opposite directions, they cancel each other out and it's the same as no force at all. Starting now, you should think in terms of what is called **net force**. Net force means what's left over when you figure in all the effects of different forces acting on something. So if you have five people pushing on something and four of those forces cancel each other, the net force, and the only one you and Isaac Newton should worry about, is the fifth force.

Figure 1.15

In Which We Describe Motion and Then Change It

As far as inanimate objects are concerned, if you know what the object is doing and know all the things that will affect that object, you can predict what the object will be doing at a later time. That's useful for lots of things, such as flying planes, designing cars, building dams, and getting electricity to your house.

Things to do before you read the science stuff

Grab a small ball or marble and find a smooth, flat surface such as a table top or a floor (sans carpeting). Roll the ball slowly across the surface. Roll it again, a bit faster this time. Try a third time, faster yet. Now roll the ball across the surface in different directions—you know, once toward the wall, once toward the refrigerator, and once trying to nail that cockroach that's scampering across the floor.

Get a ruler or a yardstick and measure off a distance of, oh, about 1 meter. The distance really isn't important, but the numbers I use later will assume you measured 1 meter. Mark the beginning and end of this distance with masking tape. What you're going to do is time how long it takes for the ball to roll this distance. In order to do this, start rolling the ball *before* the first mark, start timing when it reaches the first mark, and stop timing when the ball reaches the second mark.

Figure 2.1

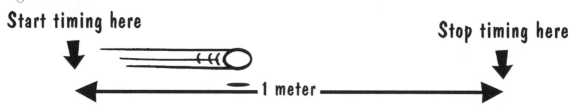

Start timing here

Stop timing here

——1 meter——

For timing things, a stopwatch would be nice but it's not necessary. A clock or watch with a second hand will do or just counting seconds (one one thousand, two one thousand, ...) works just fine. Hey, this isn't rocket science; that comes later.

Once you're all set, go for it. Time a few runs. If you're not using a stopwatch, try to guess at half seconds and such. If you come up with times that are less than a tenth of a second or more than four seconds, you're doing something weird. Check your numbers to make sure that rolling the ball faster results in shorter times and rolling the ball slower results in longer times.

The science stuff

Before you predict what an object is *going* to do, you have to know what it's doing right now. So it's kinda nice to know where the object is, how fast it's moving, and in what direction it's moving. The "how fast" part is covered by the object's **speed**. Speed is pretty much what you think it is—how far something goes in a given amount of time. Cars travel at speeds from 0 to 120 miles per hour (race car drivers and 16-year-olds excluded), planes travel from 0 to 700 miles per hour, sound travels at about 340 meters per second (around 1100 feet per second), and snails travel at about 1 centimeter per minute (that is, if they're excited). To figure out the speed of something, just divide the distance it travels by the time it takes to go that distance.

$$\text{speed} = (\text{distance traveled})/(\text{time to travel that distance})$$

Note for the math phobic

Just so you don't get totally intimidated by the symbols in an equation, here's a brief explanation. An equals sign means that whatever is on the left side of the sign is numerically the same as what's on the right hand side. If two things are in parentheses and next to each other, as in (speed)(time), that means you multiply the two together. If there's a slash between those parentheses, as in (distance)/(time), you divide the first by the second. If there are just letters and no parentheses, then two letters next to each other, as in vt, should be multiplied and two letters with a slash in between, as in d/t, means divide the first by the second.

Let's assume it took your ball 2 seconds to travel 1 meter. Then the ball's speed is

$$\text{ball's speed} = (1 \text{ meter})/(2 \text{ seconds})$$

Using your handy calculator or doing (gasp!) long division, you can divide 1 by 2 and get that the ball's speed is 0.5 meters per second. If you want your speed in centimeters per second, just plug in 100 centimeters instead of 1 meter (because there are 100 centimeters in a meter):

$$\text{speed of ball} = (100 \text{ centimeters})/(2 \text{ seconds})$$

$$= 50 \text{ centimeters/second}$$

One trick that scientists use to check whether they've done things right is to see whether the answer sorta makes sense. If you get a speed like 300 meters per second (close to the speed of sound), you messed up the calculation. You would also know you goofed if you got an answer of around 0.005 meters per second, because that would be more like the speed of a snail. What we have here is not sound or a snail, but a ball rolling along the floor. It seems reasonable that a ball could travel 50 centimeters in one second.

Now when you see formulas in textbooks, they don't usually contain words such as speed of ball, distance traveled, and time of travel. More likely they're just letters. And while you might be thinking that is just "the better to confuse you with, my dear," it saves time and ink to use the letters rather than words if you're going to write them a bunch of times. Speed can be represented by s for speed, r for rate, or v for velocity (see below for the difference between speed and velocity). Distance traveled can be represented by d (makes sense), x (represents a distance along the x-axis, for you math types), or s. You'll probably find the symbol s (standing for "spatial displacement") only in older books. The upshot of all this is that you'll see the basic definition of speed written lots of ways: $s = d/t$, $r = d/t$, $v = d/t$, $v = s/t$, and so on. How to keep them straight? Don't put too much stock in relying on the letters. Make sure you know what the letters *stand for*, and you can't go wrong.

Chances are you've noticed in your life that the direction something travels is often as important as how fast it travels. Compare the price of a regular speeding ticket with one you get while going the wrong way on a one-way street. Tired of getting reckless driving citations, scientists invented the concept of **velocity**, which tells not only how fast something is going, but in what direction it's going. So the *speed* of the ball you've been rolling might be 0.5 meters per second, but its *velocity* is 0.5 meters per second north (or "south" or "east" or "thataway" or "toward the cockroach" or whatever). We could get real

SCiLINKS.
THE WORLD'S A CLICK AWAY

Topic: velocity

Go to: *www.scilinks.org*

Code: SFF03

Figure 2.2

complicated with the math by getting into ways of specifying directions using coordinates and angles and such, but lucky for you we won't. I'll just introduce a fairly easy idea, that of representing velocity with an arrow. The direction of the arrow shows which way something is moving, and the length of the arrow gives you an idea of the speed (shorter arrow—smaller speed).

These arrows have a special name; they're called **vectors**. Vectors can represent velocities, but lots of other things, too. In fact, anything that has a **magnitude** (size) and **direction** can be represented as a vector. A few things that one can use vectors to represent: forces, accelerations (see Chapter 3), and magnetic field lines (see the *Stop Faking It* book on Electricity and Magnetism). Vectors don't crop up in your normal cocktail party conversation, but the next time you're on an airplane that lets you hear the radio communications through headphones, listen in and you'll hear the control tower telling the pilot to achieve a "vector heading" of such and such. Basically, the control tower is telling the pilot which way to go and how fast to be going there.

This book won't deal a whole lot with vectors except to use them in drawings to clarify things. If you're feeling masochistic, pick up a college physics or math textbook and learn all about vector addition, subtraction, multiplication, and what not. If you can't get enough of sines and cosines and all that, you'll pretty much be in heaven. For the rest of us here on Earth, though, there are more activities to do.

More things to do before you read more science stuff

Grab that ball again and find a surface that will slow the ball gradually to a stop after you've rolled it. Carpet works well. Now you're going to do almost the same thing you did in the previous activity, which is to time how long it takes the ball to travel a certain distance. The catch this time is that you're not going to measure the distance first. Just mark a starting point and time how long it takes the ball to go from there to the point that it stops. Remember that counting is as good as a stopwatch here.

Figure 2.3

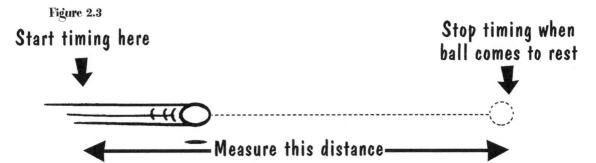

Once the ball has stopped, go ahead and measure the distance from the starting point to where the ball stopped. No points for accuracy in measuring the distance, so just go for a quick estimate. Now that you have a distance and a time, figure out the speed using speed = distance/time.

Time to take the show on the road—literally. Hop in your favorite car and find an empty parking lot or a relatively deserted street. It you can't find a place without lots of cars and people, get someone to drive you because you won't be paying full attention to the road.

Get in the car and find the odometer, which is the little counter on the dashboard that tells how far you've driven. The column on the far right, which is usually white, measures tenths of miles.[1] Just see what this reads and write it down. Now drive the car a ways, say about 6/10 (0.6) of a mile, but don't go a constant speed. Speed up to 30 miles per hour (mph), back to 10 mph, up to 40 mph, back to 5 mph, and so on. After you've gone a distance that's measurable on your car's odometer, stop and write down the total distance traveled. Oh yeah, you were supposed to be timing this, too. If you read this entire section before you got in the car, no problem.

Use the distance traveled and the time to calculate your speed for the trip. Something like

$$\text{speed} = 0.6 \text{ miles}/2 \text{ minutes}$$

$$= 0.3 \text{ miles/minute}$$

---

[1] Newer cars have digital odometers. On those, don't look for the white column, but rather the number to the right of the decimal point.

To convert this to miles per hour, just multiply by 60, because there are 60 minutes in an hour.

$$(0.3 \text{ miles/minute})(60 \text{ minutes/hour}) = 18 \text{ miles/hour}$$

If you don't understand why multiplying by 60 gets you to miles per hour, pick up a science text and look up "conversion of units." Hey, if we covered all the details here, this *would* be a science text.

More science stuff

Let's start with something obvious. The ball you rolled on the carpet was not moving at the same speed the whole time. It started out kinda fast and got slower and slower until it stopped. Yet when you calculated its speed (**speed = distance/time**) you got *one* number, maybe something like 30 centimeters (about 1 foot) per second. That number was the **average speed** of the ball during that time. At any instant, though, the ball was most likely not traveling at the average speed. For example, it might have started out at about 0.5 meters per second and it certainly ended up traveling 0 meters per second (it stopped). Providing you were watching the **speed**ometer (not the odometer) of your car as you cruised along, you could actually tell how fast you were going at any instant, and most of those readings weren't the same as the average speed for the trip (18 mph in my example). The speed you're traveling at a given instant in time is called your **instantaneous speed.** (Clever name, no?)

Now suppose you take a short trip in your car to visit dear old Aunt Edna and travel at an average speed of 50 mph for 2 hours. How far would you travel? Well, our definition of speed can be turned around with no trouble.

average speed = distance traveled/time of travel

becomes

distance traveled = (average speed)(time of travel)

which is that **distance = rate x time** formula that stumped you in fourth grade. If the little math manipulation we just did made your shorts bunch up, don't worry and just accept it for now. Anyway, we now just plug in our speed of 50 mph and time of 2 hours and get

distance traveled = (50 mph)(2 hours)

= 100 miles

Big deal. You could have done that without the formula, most likely. But the formula does come in handy when the numbers aren't so easy.

Now during your trip (which isn't over by the way—Aunt Edna lives 120 miles from your house), your instantaneous speed probably changed a lot. Sometimes you went 45 mph, other times 65 mph, and all sorts of other speeds, such as that speed of 0 mph while little Suzy was losing her cheeseburger by the side of the road. You are undoubtedly wondering about the meaning of putting an *instantaneous* speed into the formula **distance = (speed)(time)**. Even though you're not wondering that, the answer is—not much. All it would tell you is how far you *would* have traveled had you stayed at that same instantaneous speed for the time period you're interested in. "If we sit here by the side of the road forever, we'll never get to Aunt Edna's" (true). "If we travel at 65 mph for 2 hours, we'll go 130 miles and be 10 miles past Aunt Edna's" (also true). So for everyday concerns, average speed is what you should worry about.

Of course, police radar does a very good job of picking up your instantaneous speed, so you do have to worry about it once in a while.

Even more things to try before you read even more science stuff

Ball and smooth surface time again. Start the ball rolling across the surface—not too fast. Once it's rolling, try to change the speed of the ball, but not the direction it's moving. So as not to keep you in everlasting suspense, you can do this by pushing it from behind or slowing it from the front. You could also get gravity to help you by sending the ball up or down a ramp.

Get the ball moving again and this time, change the direction it's moving but not its speed (a "just right" hit from the side will do the trick). Finally, change the speed of the ball *and* the direction it's moving. A "not just right" hit from the side should do it.

Even more science stuff

If you did what I just told you to do (and you really should have if you want to learn this stuff), then each time you exerted a force on the ball, you changed the state of motion of the ball. To be more precise, you changed the *velocity* of the ball each time. Remember that velocity includes both speed and direction, so if you change the speed or the direction or both, you're changing the velocity. There's a special name for a change in velocity, and it's **acceleration**. Examples of things that are accelerating: a braking car (changing speed but not direction), a car that's speeding up (why do you think they call that pedal an accelerator?), the expensive piece of china that your kid just dropped (watch it speed

SCI**L**INKS.
THE WORLD'S A CLICK AWAY

Topic: acceleration
Go to: *www.scilinks.org*
Code: SFF04

up as it falls to the floor), a dog chasing its tail (change in direction), a roller coaster nearing the end of a vertical loop (change in direction and quite possibly a change in speed), and your lunch as you lose it after that vertical loop (change in attitude of the person next to you).

An exact definition of acceleration, if you really want one, is the change in an object's velocity divided by the time it takes to make that change.

acceleration = (change in velocity)/(time for the change)

You might see this in textbooks as

$$a = \Delta v / \Delta t$$

which makes sense only after you know that the Greek letter delta (Δ) means "change in." If you pick up a textbook that uses calculus, you'll see the same thing written as $a = dv/dt$.

If you change an object's velocity a whole lot in a short time, that's a big acceleration. If you change an object's velocity just a little over a long period of time, that's a small acceleration. For the record, acceleration is a vector because it has both magnitude (size) and direction. But if you don't want your brain to snarl up in little knots right here and now, don't think too hard about acceleration having a direction.

Anything that's moving in a straight line (no change in direction) at a constant speed (no change in speed) is *not* accelerating. A trivial case of moving at a constant speed is no speed at all, so it's safe to say that as you park yourself on the couch and watch reruns of *Andy Griffith,* you're not accelerating. [Note to the nitpickers: Yes, couch potatoes really do accelerate because they turn in circles (change in direction) along with the spinning Earth, but I'm trying to keep things simple.]

For a nontrivial example of nonacceleration, close your eyes and imagine you're cruising along in a jet at 33,000 feet with no turbulence. Nice and smooth. You can almost taste the snack of the day (What? Pretzels again?). You can travel at 500 mph and not accelerate as long as there's no change in speed or direction. Then WHAM! You hit a big air pocket and fall 500 feet in 2 seconds. That counts as a big ol' acceleration and, more importantly, you can *feel it.* Which is the point I'm trying to make. You can feel accelerations. Back in the plane, you can notice even the slightest turbulence with your eyes closed because each little bump is a change in speed and/or direction. You can feel it when a car rounds a curve. You can feel it when an elevator starts and stops (both accelerations), but you don't notice the motion in between (no acceleration).

Way back at the beginning of this chapter, I talked about predicting where something will be and what it will be doing at a later time *if* you know where it is now, what it's doing now, and what affects it in the near future. What that amounts to is knowing the object's position (where it is), velocity (what it's doing), and acceleration (how its velocity is changing). Well, there are some fancy formulas that you can just plug into and get any information you want, and since I'd hate to send you on your way without some stuff to impress friends and family, I've written those formulas below. I don't expect you to fully understand what all the letters and subscripts mean—no sense going into all that unless you're actually going to use them to solve problems, which we are not going to do.

(where object is after time t) = (where it is now) +
(speed it's moving at now)(time t) +
1/2(acceleration of object during time t)(time t)^2

or

$$x = x_0 + v_0 t + 1/2 a t^2$$

(velocity of object after time t) = (velocity of object now) +
(acceleration of object during time t)(time t)

or

$$v = v_0 + at$$

If the acceleration during the time t changes, then these formulas don't work, but they're pretty useful. Go find a high school or college textbook and look under the chapter on "kinematics." You'll see our friendly formulas there. Of course, don't expect that you know everything there is to know about these formulas, because it does take a wee bit of practice to use these correctly. But at least you'll have a feel for what the letters stand for and what the formulas are used for.

Chapter Summary

- The speed of an object is defined as the distance it travels divided by the time it takes to travel that distance.

- When you specify the direction an object is moving in addition to its speed, you are describing the object's velocity.

- Instantaneous speed and velocity are, in general, different from the average speed and velocity of an object. When describing an object's motion, it is important to distinguish between these quantities.

- Any change in an object's velocity, which can be a change in speed or direction or both, is known as an acceleration.

- The human body can detect when it is accelerating.

- There are a number of mathematical relationships between distance, velocity, acceleration, and time that are useful for describing an object's motion, determining is past motion, and predicting its future motion.

Applications

1. Remember Rover from the **Applications** section of the first chapter? He was flying through the air in your car after you slammed on the brakes. As he flew through the air, foremost in your mind had to be the question, "When was he accelerating and when not?" When you slammed on the brakes, Rover was the only one who *didn't* accelerate. The car accelerated because it changed speed. By virtue of your seat belt, which exerted a force on you, you also accelerated. By the way, it doesn't matter whether you speed up or slow down; it's always called acceleration. And yes, that is a bit confusing. Now Rover just kept on doing whatever he was doing—his velocity stayed the same. Until he hit the dash, of course.

2. To take an example from the classics, let's talk about the tortoise and the hare. We all know that the tortoise won the race, but what about that all-important comparison of the average and instantaneous velocities of the two? Well, average velocity is just the total distance divided by the time, and since the tortoise won the race, his average velocity was greater than the hare's. Because the tortoise probably plodded along at pretty much the same speed throughout the race, his instantaneous velocity at any time was probably the same as his average velocity for the entire race. The hare's instantaneous velocity was either zero (taking a nap) or pretty fast (when he was running). So we all know that the moral of the story is that instantaneous velocity don't mean nuthin' unless you win the race.

Figure 2.4

Large instantaneous velocity   Instantaneous velocity zero

3. Amusement parks are for being amused, but different ages get amusement out of different things. Little kids are into no accelerations (rides that move in a straight line at a constant speed) or small accelerations (those little boats that very *slowly* change their direction). Big kids like big accelerations, such as quick changes of direction and quick changes in speed. Rides that twist and turn and go alternately fast and slow fit the bill. Of course, this all ties into the fact that you can *feel* accelerations.

Topic: roller coaster physics

Go to: *www.scilinks.org*

Code: SFF05

Newton's Second One

I f you understood the first two chapters, you now know how to describe motion (using velocities and accelerations) and you know what causes changes in motion (*net* forces do it). Remember that net force refers to what you get when you consider the total effect of all the forces acting on something. If two equal forces are acting in opposite directions, the net force is zero. And of course you remember that an acceleration is any change in speed and/or direction. So a net force acting on an object causes the object to accelerate. All this knowledge makes you a junior science nerd, so why not try for apprentice nerd by learning exactly how forces and accelerations are related.

Newton's 2nd

Things to do before you read the science stuff

Still have that marble or ball or whatever it was you used in the last chapter? If not, get another one. Now see if you can find a ball that's a lot heavier or lighter than the one you're using. A marble (light) and a baseball (heavy) make a good pair; so do a small and large ball bearing. Size doesn't matter a whole lot, as long as the two balls have noticeably different weights.

Place the balls side by side on a smooth surface and check out how hard you have to push each one to get it moving at a given speed. You should have to push the heavier one harder. If you ignored my advice and the two balls are pretty close to the same weight and you can't tell the difference in how hard you have to push them, go outside and try pushing your car to get it moving the same speed as the balls. Don't send me your medical bills if it doesn't occur to you to put the car in neutral, release the parking brake, and make sure you're not pushing uphill.

Assuming you didn't seriously hurt yourself with the car, put the heavy and light balls back on the smooth surface. Try to push them *equally hard* and see what happens. Okay, okay, it's not easy to push them equally hard, but you can probably come close. If your pushes aren't totally messed up, the push should have a greater effect on the light ball than on the heavy ball. The heavier the object, the less effect your push should have (try that car again if you need convincing).

Time to do that old "hit from the side" thing again. Roll one of the balls along the surface and whack it from the side with varying forces (Figure 3.1). In

Figure 3.1

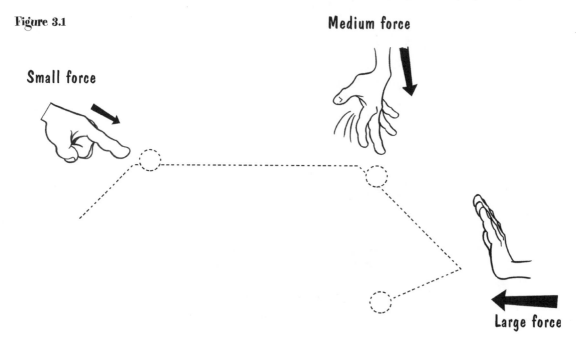

Small force

Medium force

Large force

other words, hit it just a little, more than that, and then a lot. Notice any difference in how the ball reacts to different-sized forces.

The science stuff

In case you were sleeping through Chapter 1, I'll remind you that it's hard to change the motion of an object that has lots of inertia and it's easy to change the motion of an object that has little inertia. No big surprise then that the heavy ball you just played around with has more inertia than the light ball. Since we've gone several pages without a new vocabulary word, here's one: **mass**. Mass is the measure of an object's inertia. Objects with little inertia have a small mass and objects with a lot of inertia have a large mass. In fact, anything you can say about an object's inertia can be said about its mass, because they're pretty much the same thing.[1]

Topic: mass

Go to: *www.scilinks.org*

Code: SFF06

Now, remembering that a change in speed and/or direction is an acceleration—and that all pushes, pulls, nudges, hits, and bumps are called forces—see if the following doesn't agree with what you found as you did the last activity section.

Applying a larger net force to an object results in a larger acceleration (Figure 3.2).

Figure 3.2

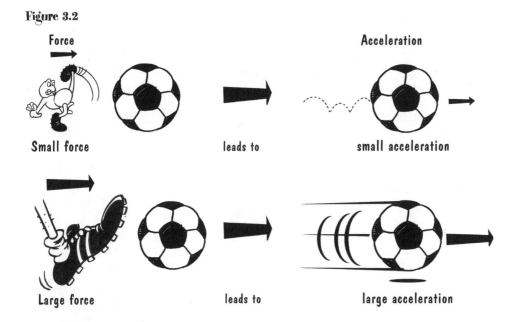

---

[1] If this book inspires you to study physics as a hobby, you'll eventually find out that there are two different definitions of mass—inertial mass and gravitational mass. That distinction is beyond the scope of this book, so we won't get into it.

If you apply equal net forces to two objects, the one with the smaller mass will accelerate more (Figure 3.3).

Figure 3.3

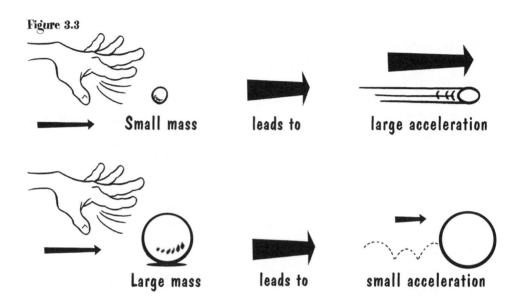

Small mass   leads to   large acceleration

Large mass   leads to   small acceleration

If you want to cause two different objects to have the same acceleration, the object with the larger mass will require a larger force (Figure 3.4).

Figure 3.4

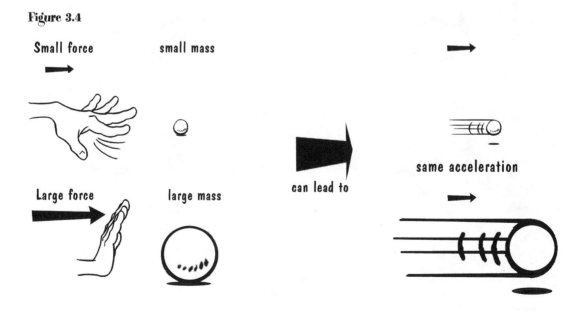

Small force   small mass

Large force   large mass   can lead to   same acceleration

Remember that I'm using the term *net* force because if the forces you apply to an object cancel each other out, you're not going to accelerate the object. (Think back to the movers and the couch in the **Applications** section of Chapter 1.) Also, accelerations don't have to involve things speeding up or slowing down. When you hit the ball from the side with different-sized forces, maybe all that happened was the ball made a sharper change in direction each time. Everything else being equal, a large change in direction is a bigger acceleration than a small change in direction (Figure 3.5).

Now if you were writing a book about forces and accelerations, you might want a shorthand way to state everything that I just stated above. First, I'll write it in words:

Net force acting on an object = (mass of the object)(acceleration of the object)

In symbols, it looks like this:

$$F = ma$$

Those three little letters, plus the equals sign, are a statement of **Newton's second law**. What I'm going to do is put some numbers in for *F*, *m*, and *a* in order to show how this relationship describes the results you got with rolling balls and such. What I don't want you to do is worry about how one decides that a force has a value of 10 or a mass has a value of 2 and so on. Just concentrate on the overall relationship. Any physicist who's reading this right now is cringing because the numbers should also have *units* attached to them—things like kilograms and pounds and such. But that would get in the way of your understanding, so go ahead and let them cringe.

Anyway, suppose you're pushing on a ball that has a mass whose value is 5. The first time you push, you push with a force of 10, and the second time you push, you push with a force of 20 (twice as big a push). If you put a force of 10 and a mass of 5 into Newton's second law, you get

$$F = ma$$

$$10 = (5)(?)$$

Figure 3.5

Small change in direction—small acceleration

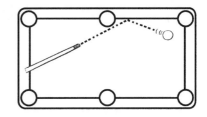

Large change in direction—large acceleration

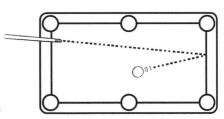

Since the left hand side has to equal the right hand side, the *a* (acceleration) must equal 2, because 5 times 2 is 10. So a force of 10 applied to a mass of 5 gives an acceleration of 2.

Now change the force to 20. The mass is still 5.

$$F = ma$$

$$20 = (5)(?)$$

Because 5 times 4 gives you 20, the acceleration is now 4. Twice as big a force (20 instead of 10) results in twice as big an acceleration (4 instead of 2). The numbers say the same thing you already knew about hitting an object with a larger force. Now that you're energized with the number thing, let's see what *F = ma* says about applying the same force to different masses. Suppose you use a force of 10 to hit two different balls, one of mass 5 and another of mass 2. For the ball of mass 5, you get

$$F = ma$$

$$10 = (5)(?)$$

This is the same as the first situation, and you get an acceleration of 2. Now do the same for the ball of mass 2.

$$F = ma$$

$$10 = (2)(?)$$

For the stuff on the left to equal the stuff on the right, the acceleration has to be 5. Once again, the numbers tell you what you already knew. Applying the same force to two different masses causes the smaller mass to have a larger acceleration.

All right, you can unpucker your mouth now. No more math for a while. In fact, I'll give you a visual tool for understanding Newton's second law. Think of the *F* and the *ma* as being on opposite sides of a teeter-totter that you always want to keep in balance. (See Figure 3.6.)

If you make *F* larger or smaller, then *m* or *a* or both of them will have to change such that *m* times *a* gets larger or smaller also (the situation will tell you which changes). If *F* stays the same and *m* gets larger, *a* will obviously have to get smaller for things to stay in balance. If *m* gets smaller, then *a* has to get larger.

SCI LINKS.
THE WORLD'S A CLICK AWAY

Topic: Newton's second law

Go to: *www.scilinks.org*

Code: SFF07

Figure 3.6

Before going on, I should tell you that Newton's second law isn't always as easy as figuring out teeter-totters and equations with simple numbers. Things get messy when you try to figure out what happens when the acceleration involves a change in direction as well as a change in magnitude, and it's not always an easy thing to figure out what number to put in for *F*. But we *are* sticking to the basics here, aren't we?

One thing you should know about science, though. When you try to apply it to simple, everyday situations, things can get complicated really fast. Trying to figure out all the forces that act on a paper bag blowing in the wind and then figuring out its acceleration and then figuring out where it will end up is pretty much an impossible task.

More things to do before you read more science stuff

Scrounge around the house for a long piece of string (at least 3 meters), a plastic straw, some masking tape, two large paper clips, a 5 x 7 index card, and two balloons (sort of hot dog shaped are the best, but any will do). Also see if you can find some semi-heavy objects that you can hang from a hook. Metal washers are great, but even small plastic cups with handles work. Coffee cups are probably too heavy. In what follows, I'll assume you found some washers.

Unwrap the straw and tape it along one edge of the index card. Also form the paper clips into hooks and punch them through the index card as shown in Figure 3.7.

Tie one end of the string to something like a door handle and stretch the string across the room. Thread the straw onto the free end of the string. Blow up one of the balloons but

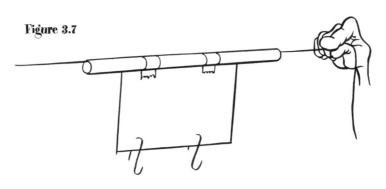

Figure 3.7

don't tie it off. While holding on to the end to keep air from escaping, tape the balloon to the index card (Figure 3.8). If manual dexterity isn't your thing, get somebody to help you do this.

Hold the string tight and release the balloon. It should zoom to the other end of the string. If it hit you in the face, try again with the bal-loon aiming in the other direction.

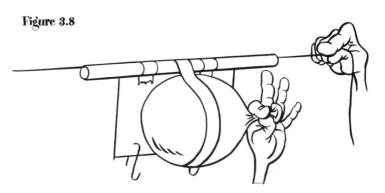

Figure 3.8

Now that you know one way to entertain the kids at that next birthday party, mess around with the setup. That means try adding washers to the paper-clip hooks (this adds mass to the card) and try using more than one balloon (this increases the force that pushes things along). Basically, you're changing force or mass or both and seeing what effect that has on the acceleration. Use your $F = ma$ teeter-totter to explain what's going on. For a real challenge, try changing the position of the string so the balloon rocket angles upward instead of going horizontal. Explain your result if you can.

More science stuff

Actually, I'm not going to give you anything new here. Just a check to see if you understood how Newton's second law applies to the balloon rocket. Hopefully, the acceleration (measured by how fast the rocket got moving) increased when you increased the force (added balloons) and decreased when you increased the mass. In terms of the teeter-totter, increasing F means the other side of the teeter-totter has to get larger also. If you keep the mass the same, the only choice is that the acceleration increases. If you keep F the same (keep using only one balloon) and increase the mass (add washers), then the acceleration has to get smaller to keep the teeter-totter in balance.

When you angled the string upwards, the acceleration got smaller, right? But why? Did the mass get larger? Nope. Did the force get smaller? Yep. Remember that F stands for the *net* force. You have to figure in *all* the forces acting on the rocket. With the rocket horizontal, gravity pulls on it, but that doesn't affect things because it's not pulling in the direction of motion. But when the string is angled upward, gravity is partially pulling back on the rocket. So the net force (force caused by air escaping the balloon minus a component of gravity) is smaller than before and the acceleration is less (Figure 3.9).

Figure 3.9

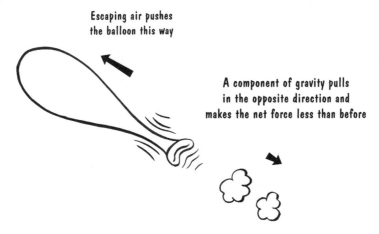

Escaping air pushes
the balloon this way

A component of gravity pulls
in the opposite direction and
makes the net force less than before

SC*i*LINKS.
THE WORLD'S A CLICK AWAY

Speaking of net forces, there's another force we haven't considered. As it moves along, the straw rubs against the string, resulting in a force of **friction**. This friction force opposes the force caused by the air escaping the balloon. So the net force is really the force caused by escaping air minus the

Topic: friction

Go to: *www.scilinks.org*

Code: SFF08

Figure 3.10

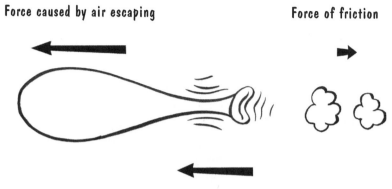

Force caused by air escaping

Force of friction

Net force is the force caused by the air
minus the force of the friction

force of friction (Figure 3.10). If you have some fishing line around the house, substitute this for the string and repeat your little rocket experiment. The friction force between fishing line and the straw is less than that between string and the straw, so the net force is larger and the acceleration should be larger.

Chapter Summary

- An object's mass is a measure if its inertia.

- A net force acting on an object causes the object to accelerate.

- The acceleration that results from a given applied force depends on the mass of the object being accelerated.

- The relationship between net force, mass, and acceleration is represented by Newton's second law, which looks like this in equation form: $F = ma$.

- When using Newton's second law to determine an object's acceleration, you have to be sure to include all the forces, including friction, which might be acting on the object.

Applications

1. My dad always said that I shouldn't follow too closely behind motorcycles because they stop faster than cars. Was he right? Yes, for once in his life he was right. (Dads are seldom right about anything when you're a teenager.) Motorcycles can also get going from a stop faster than cars can (we're talking garden-variety cars and motorcycles here, not the kinds that spit fire).

The reason is that motorcycles have a smaller mass than cars. In general, it's easier to accelerate them than it is cars (remember that both stopping and starting are called accelerations). Of course, motorcycles also have smaller engines that don't deliver as much force, but the smaller mass is the dominant thing here. If you understand how all this works, then go ahead and explain why you should be doubly worried when an eighteen-wheeler is riding your butt.

2. As a car rounds a curve (we're on a car theme here), it's accelerating because it's changing direction. The force that's causing this acceleration is friction between the tires and the road. First, notice in Figure 3.11 the direction of the friction force—toward the inside of the curve. Why is that? Not obvious at this point, and I'm just going to ask you to trust me on this one until you get to Chapter 6. Assuming I have the direction of the friction force correct, what kind of curves—sharp ones or gradual ones—will wear out your tires faster?

Figure 3.11

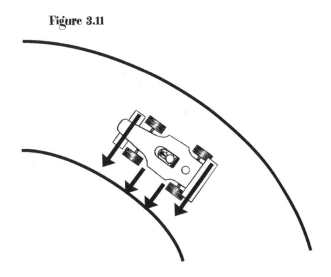

Gradual curve= small friction force

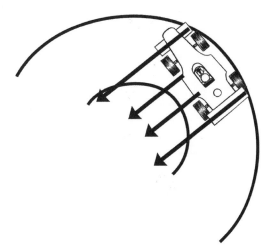

Sharp curve= large friction force

The commonsense answer is that sharp curves wear out your tires faster, but isn't it nice to know that Newton's second law gives the same answer? Sure, it is. In a sharp curve, you're changing your velocity faster than you are in a gradual curve, because you're changing direction faster. A faster change in velocity is a larger acceleration, and larger accelerations require larger forces. That means the friction between your tires and the road is a larger force in sharp curves (Figure 3.11). A larger friction force wears your tires out faster.

3. In Figure 3.12, determine the mass m that will allow the 100 kg block to accelerate down the frictionless incline at 0.3 m/s². Assume that the pulley has no friction and that $g = 10$ m/s². Ha, ha, just kidding. Thought you'd like to know what you were missing by not working through a college text.

Figure 3.12

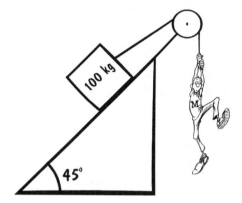

There's No Such Thing as Gravity–The Earth Sucks

The title of this chapter is a stupid joke you learn early on if you ever take a physics course, and it's best appreciated in adolescence, arrested or otherwise. Before starting, let me use gravity to illustrate a point about science. In doing workshops for teachers, I often drop something and then ask the teachers to tell me why the object fell. The answer is "gravity." I then ask what gravity is. After some hemming and hawing, we all arrive at an answer something like "Gravity is the thing that makes objects fall to the Earth." So, objects fall to Earth because of gravity and gravity is the thing that makes objects fall to Earth. See any circular reasoning there? The point is that you haven't explained *why* objects fall; you've just given it a name. After reading through this chapter, you should know more about gravity than you did before but you still won't know *why* objects fall to Earth. They do it because that's the way things are. If you want to know why things happen, consult your favorite religion, because science never does answer the "why" question. This doesn't account for people whose religion *is* science, but I'll save that for another book.

Before going on, I'll expand just a bit on the "why" issue. When I say that science doesn't answer the "why" question, I mean in an ultimate, grand-scheme-of-things sense. In a *practical* sense, of course science explains things. When you ask "Why is the sky blue?" there's a scientific answer (see the *Stop Faking It* book on Light). To go back to the gravity example, you could ask why things have a gravitational attraction for one another. There are scientific answers for that. A learned person, in answer to your question, might launch into an explanation involving "gravitons" or the "curvature of space-time." But then you could ask why gravitons exist or why space-time is curved. Depending on the person's knowledge of physics, he or she could possibly take you to another level of explanation, but at some point, when you ask why the latest explanation is the correct one, your inquisitee will have to resort to something such as "Because that's just the way things seem to be." Either that, or the person will give you an answer that is religious in origin. Scientific explanations are powerful, but they have their limits.

Figure 4.1

Things to do before you read the science stuff

Get the light and heavy balls you used in the previous chapter. Also see if you can find a box of paper clips or thumbtacks. Go outside (or stay indoors if you have carpeting) and drop the heavy and light balls side by side, from the same height, at the same time. Which one hits the ground first? Now take one paper clip out of the box and answer the following question: which has more mass, a single paper clip or a box of paper clips? If you answered "the single paper clip," go get a cup of coffee and try again. Drop the single paper clip and the box of paper clips (might be a good idea to close the box) side by side, from the same height, at the same time. Which hits the ground first? You might have to repeat this little experiment a few times before you decide.

Throw a ball straight up into the air and watch what it does. Describe its motion. You know, where does it slow down, speed up, stop, and all that? Is the ball accelerating? If you think the ball is accelerating, decide what force or forces are acting on the ball in order to accelerate it. (Quick review: an acceleration is any change in speed and/ or direction, and accelerations are caused by forces.)

The science stuff

Topic: gravity

Go to: www.scilinks.org

Code: SFF09

Just in case you haven't gotten that coffee yet, I'll remind you that when you drop something it falls to the ground. Though it might not be obvious, the object speeds up as it falls. That means it's accelerating (change in speed) and there must be a force acting on it. The force that's acting on it is—all together now—**gravity**. I'll discuss gravity that has nothing to do with the Earth later, but for now let's just say that gravity is the Earth's pull on things.

What did you decide about the ball you threw in the air? Accelerating? If it was changing its state of motion, then it was accelerating. Well, it goes slower and slower till it stops at its highest point, and then it speeds up as it heads to the ground. Sounds like a change in motion to me, so it was accelerating. What force caused the acceleration? It's that invisible force of gravity. In fact, if you ignore air friction (scientists always ignore real-world complications to make a problem simpler!), gravity is the *only* force acting on the ball. Anything that's in free-fall has gravity as the only force acting (again, ignoring air friction).

Figure 4.2

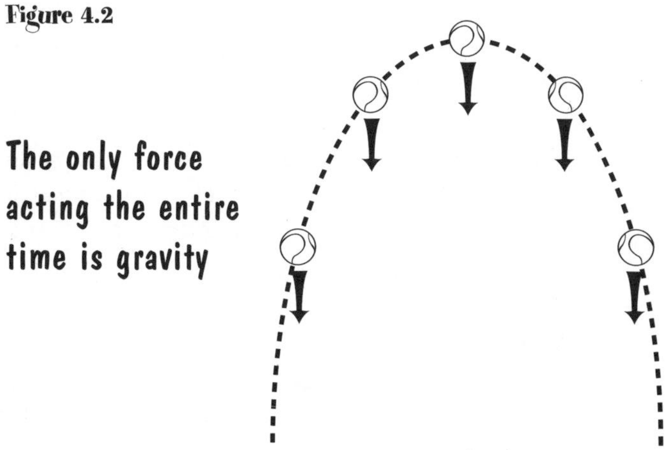

The only force acting the entire time is gravity

Now if there's a net force acting on something and it's accelerating, that means our good buddy, Newton's second law ($F = ma$), ought to describe what's going on. I'll write this relationship for a single paper clip and a box of paper clips, both falling under the influence of gravity.

(Force of gravity   = (mass of single clip)(acceleration of single clip)
    acting on single clip)

and

(Force of gravity   = (mass of box of clips)(acceleration of box of clips)
    acting on box of clips)

First pay attention to the two forces that are on the left hand sides of the equations. The force of gravity acting on the single paper clip is much less than the force of gravity acting on the box of paper clips. How do you know? You can *feel* the difference. Holding them side by side, you can feel that gravity is pulling harder on the whole box than on the single paper clip. Of course, you already knew that, because you know that the whole box is *heavier* than the single paper clip, or that the **weight** of the box is greater than the weight of the single clip. And by golly, "weight" is the term scientist folks use for the force that gravity exerts on something.

Because you read and thoroughly understood Chapter 3, you also know that the mass of the single paper clip is less than the mass of the box of paper clips. It's easier to change the motion of (accelerate) one paper clip than it is to change the motion of (accelerate) a box of paper clips (try flicking each with your finger).

Okay, what accelerations do we use for the right hand sides of the two $F = ma$ equations above? I just knew you were dying to know that. Hopefully, you found that when you dropped the box and the single paper clip side by side, they hit the ground at about the same time. Same for the little ball and the big ball. (Remember, we're not talking exact results here because you didn't use some electric-release doohickey. If the two objects hit pretty close to the same time, that's good enough.) Assuming they did hit about the same time, that means they had the same acceleration, because both started at zero velocity and changed their motion in the same way so as to hit at the same time.

So what we have is a couple of balanced $F = ma$ teeter-totters that look like this:

Figure 4.3

National Science Teachers Association

Apparently, then, here's how gravity works. The more mass an object has, the larger force gravity exerts on it. But it's also harder to accelerate an object with a larger mass. Gravity pulls harder on objects with a larger mass, but the larger mass makes it so the acceleration remains the same. In fact, **when gravity is the only force acting, all objects have the same acceleration**. I'll show you later that this isn't just a coincidence. In textbooks, you'll see the acceleration due to gravity written as g, and the value of g is 9.8 meters per second per second (32 feet per second per second).[1] Don't worry, I won't make you get out your calculator.

Now that you know how gravity affects objects, it's time to discuss the often-misunderstood difference between *mass* and *weight*. It's pretty easy, really. Mass is a measure of an object's inertia—how hard it is to change the motion of an object. Mass is what goes in for the m in $F = ma$. Weight is the *force* that gravity exerts on an object. $F = ma$ tells us that forces and masses are different things. If you were able to turn off the force of gravity tomorrow, you would still have the same mass—it would be just as difficult to change your motion as before. But your weight would now be zero because the Earth's gravity would no longer be pulling on you. Zero weight but your thighs still rub together—where's the fun in that?

The reason people get so confused by the mass and weight thing is that the two are so closely related. If you pig out on hot fudge sundaes for two weeks, your mass will increase. The increase in mass also causes your weight to increase, so it's natural to think of the two as being the same. Of course, those bathroom scales that measure your weight in pounds (an appropriate unit for measuring weight) and also give a readout in kilograms (a unit of mass, not weight) don't help any. Suffice to say that if you don't understand Newton's second law, you'll have a difficult time understanding the difference between mass and weight.

If you *do* understand Newton's second law, then you're in business. Weight is a force, and thus belongs on the F side of the equation. Mass is what goes in for the m on the right side of the equation. Weight is how hard the Earth pulls on something. Mass is how much of the something there is, and is a measure of how hard it is to change the state of motion of that something. Suppose you send an object light years out into space so it's a long, long way from the Earth or any other planet. There, the object's weight is practically nothing, because the gravitational pull of the Earth gets weaker and weaker the farther you are from the Earth. The object's mass, however, is exactly the same as it was on Earth. It's just as

---

[1] The unit of "per second per second" might seem a bit strange. Remember that velocity is measured in meters per second. Acceleration is a change in velocity, so its units are "change in velocity per second." Because velocity already has "per second" in it, the unit of acceleration contains an additional "per second."

difficult to change that object's motion out in space as it is on Earth. In other words, you wouldn't have an easier time pushing an elephant in space than you would on Earth, even though the elephant's weight would be practically zero!

More things to do before you read more science stuff

The only piece of equipment you need for this section is your brain, hopefully in relatively good working order. See if Grandma and Grandpa can take the kids for a day, take a short nap, and you'll be set. What I want you to do is simply think about the questions below. Really try to answer them before you go to the next section.

1. What causes the Earth to pull everything to it? What actually causes gravity?

2. Can distant objects in the heavens influence our lives? Should you call the Psychic Hotline *today*?

3. Why does everything fall to Earth at the same rate?

4. Are astronauts really weightless when they orbit the Earth? If not, why do they need those special toilets?

5. How does the Moon cause tides? Isn't it supposed to have something to do with gravity? How many high tides are there in 24 hours? (You can look this one up.) Does the Moon have to be full to cause high tides?

More Science Stuff

1. What causes the Earth to pull everything to it? What actually causes gravity?

The short answer to these questions is that we don't know. Not all scientists will give you that answer, though. Some might say that mass causes gravity or that gravitons cause gravity or that the mass-influenced curvature of space-time (!) causes gravity. But then you can ask what causes mass or what causes gravitons or what causes curved space-time, and you eventually get back to the fact that science doesn't answer the *why* question, as I already told you in the intro to this chapter. Once again, it's religion time.

What we *do* know is that everything that has mass exerts a gravitational force on everything else that has mass. So there's a gravitational force between the lamp and the chair, between a pencil and a house, between a car and a bird, and between you and that last piece of chocolate cake in the fridge. All right then, if everything is attracted to everything else, why doesn't everything in the world just glom together in one big pile? Why don't you "fall" toward a house or a car even when you haven't had one too many margaritas? The answer is that gravitational forces between things are very, very, very, very, very, very, very small. That is, unless one of the things has a very, very, very, very, very, very, very *large* mass. It turns out that the Earth has a very, very, very, very, very, very, very large mass, so the gravitational attraction between the Earth and everything on it is sizable. The gravitational attraction between objects such as you and a car is so small as to not even be noticeable. In fact, you have to set up some sophisticated equipment to measure the gravitational force between ordinary objects. This suggests that the attraction between you and the last piece of chocolate cake must not be primarily gravitational.

A cool thing about gravitational forces is that they act *through* other materials. When you place a table between a book and the Earth, the Earth's gravity still causes the book to fall toward the Earth.

The gravitational force between two objects also gets weaker as the objects get farther apart. This is a good thing. If gravity depended only on the mass of the objects and not on the distance between them, then the gravitational pull of the Sun and other stars, which have a much greater mass than the Earth, would be much stronger than the gravitational pull of the Earth. It's not entirely clear what that would

Figure 4.4

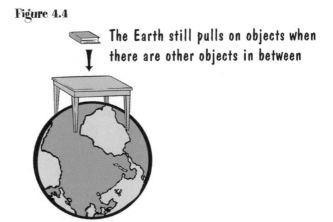

The Earth still pulls on objects when there are other objects in between

mean for your everyday existence, but at a minimum there wouldn't be any tides (step ahead a few pages to find out about tides) and our orbit around the Sun wouldn't be stable, meaning we would probably spiral into the Sun or out away from it. The unstable orbit possibility, by the way, would cut your existence short enough that you wouldn't have much time to ponder your fate. For that matter, if gravitational forces didn't get weaker with distance, the universe probably would have collapsed on itself a while ago, and you wouldn't be reading this.

I know you're just dying for a formula that sums up all of this gravity stuff, so there's one below. If you don't remember what the lines and parentheses mean, refer back to Chapter 2.

$$\text{Gravitational force between two objects} = \frac{\text{(a very tiny number)(mass of object 1)(mass of object 2)}}{\text{(distance between the centers of the two objects)}^2}$$

or, in short hand,

$$F_{grav} = G\frac{m_1 m_2}{r^2} \qquad \textit{where G is that very tiny number}$$

This formula technically only applies to spherical objects or objects with no size (yeah—uh huh), but that technicality won't bother us. We'll ignore it and it won't really affect what we're doing here.[2] If the two masses, represented by m_1 and m_2, aren't very large, the fact that G (known as the Universal Gravitational Constant) is such a very tiny number means the gravitational force is also a very tiny number. Also, if the two masses are a long distance apart, then r is a very large number and r^2 is an even larger number. When you divide by a very large number you end up with a very small number. Translation: large distances and small masses result in very small gravitational forces.

Just for kicks, let's figure out what the formula says about the gravitational force between the Earth and a typical person. It will involve putting numbers in for the symbols in the gravity formula, and it might look a little scary. If it scares you, you have two choices. One is to ignore the formulas and just pay attention to the final results. The other is to follow along slowly and pay attention to the order of things. For example, there's a G on the right side of the equation below, and it's the first term on the top. The number for G is 6.67×10^{-11}, so that should be the first term on the top of the right hand side when we put numbers in. At any rate, say this person has a weight of 150 pounds. Now, if you'll trust that I'll put in the right numbers (I'm going to use metric units, where force is measured in newtons rather than pounds, so don't pay too close attention to the numbers), here's what we get:

$$F_{\substack{\text{between} \\ \text{Earth and} \\ \text{person}}} = \frac{G(\text{mass of Earth})(\text{mass of person})}{\left(\substack{\text{distance between the} \\ \text{centers of the two objects}}\right)^2}$$

$$= \frac{(6.67 \times 10^{-11}\, m^3/kg\text{-}sec^2)(5.98 \times 10^{24}\, kg)(68.2\ kg)}{(6.37 \times 10^6\, m)^2}$$

We're assuming the person is spherical, which might not be a bad assumption in some cases. Also, the number for the mass of the person isn't the same as his or her weight, because mass and weight are different things. The distance

---

[2] Physicists are always making simplifying assumptions about the real world because the complications of the real world make it difficult to solve problems. There's an old joke about a mathematician, a chemist, and a physicist setting out to milk a cow. I don't remember most of the joke, but the punch line is that it's the physicist's turn and he begins by saying, "Let's assume we have a spherical cow." Maybe you had to be there.

between the center of the person and the center of the Earth is just the radius of the Earth. Run this thing through the calculator and you get

Force of gravity acting on person = 670 newtons, which is approximately equal to 150 pounds

Surprise! Remember that a person's weight is just the gravitational force that the Earth exerts on him or her. So maybe that isn't a surprise. By the way, the reason the calculation doesn't work out to exactly 150 pounds is that I used approximations for a couple of the numbers involved.

Okay, let's compare this force with the gravitational force between that same person and 5 kilograms (roughly 10 pounds) of chocolate fudge sitting in front of the person.

$$F_{\substack{\text{between} \\ \text{person and} \\ \text{10lbs of fudge}}} = \frac{G(\text{mass of piece of fudge})(\text{mass of person})}{\left(\substack{\text{distance between} \\ \text{person and fudge}}\right)^2}$$

$$= \frac{(6.67 \times 10^{-11}\, m^3/kg\text{-}sec^2)(5\,kg)(68.2\ kg)}{(.1m)^2}$$

$$= .0000023 \text{ newtons}$$

$$= .00000052 \text{ lbs!}$$

Small force, yes? It's almost 300 million times smaller than the person's weight. No wonder we don't notice gravitational forces between people and objects on the Earth.

2. Can distant objects in the heavens influence our daily lives? Should you call the Psychic Hotline *today*?

If astrology works for you, go for it. But in the meantime, let's see what scientific effect say, Jupiter, might have on that 150–pound person. There are four known forces (stress "known") in the universe, and as far as we know, gravity is the only one that can reach as far as Jupiter. (Electrical forces can reach that far, but things tend to be electrically balanced in the universe, so that force is irrelevant when talking about Jupiter.) So let's say for the heck of it that if Jupiter is going to influence your life, it has to do it through gravity. Jupiter's a bigger and more massive planet than the Earth, so maybe it has a big effect on you.

$$F_{\substack{between \\ person\ and \\ Jupiter}} = \frac{G(\text{mass of Jupiter})(\text{mass of person})}{\left(\begin{array}{c} \text{distance between} \\ \text{person and Jupiter} \end{array}\right)}$$

$$= \frac{(6.67 \times 10^{-11}\,\text{m}^3/\text{kg-sec}^2)(1.9 \times 10^{27}\,\text{kg})(68.2\,\text{kg})}{(6.3 \times 10^{11}\,\text{m})^2}$$

$$= .0000022 \text{ newtons}$$

$$= .00000049 \text{ lbs!}$$

where 6.3 x 10^{11}m is the closest Earth and Jupiter get to each other

What makes this such a small number is the fact that the distance between a person and Jupiter is so great: *r* being very large makes the whole force very small.

So comparing the two numbers we find that people are slightly more attracted to a bunch of chocolate fudge than they are to astrology. Not a big surprise there. But this also shows that, when you're close to something, your gravitational attraction to anything more massive than 5 kilograms (which includes a whole lot of things in the world) is also stronger than your gravitational attraction to Jupiter. Now astrologers, being people who want a lot of your money, will be quick to tell you that gravity isn't the only force between you and the heavens. Instead, it's one of those forces that scientists haven't discovered and probably a force that scientists will *never* discover. As the Church Lady might say, "How conveeeeeeenient."[3]

3. Why does everything fall to Earth at the same rate?

Because that's the way the world is. Haven't you been paying attention when I keep telling you that science doesn't answer any *why* questions? Okay, I'll give you the science answer. It'll take just a teensy bit of math, but no calculators. You recall Newton's second law, right? All together now—*F* = *ma*. For the force *F*, we're going to use that complicated expression for the gravitational force between two objects. One object is the Earth, and the other object is something, such as a box of paper clips, that's just above the surface of the Earth. So the gravitational force between the two is

$$F_{\substack{between \\ Earth\ and\ a \\ box\ of\ paper\ clips}} = \frac{G(\text{mass of Earth})(\text{mass of box of paper clips})}{(\text{radius of Earth})^2}$$

---

[3] For those of you who don't know who the Church Lady is, check out some reruns of *Saturday Night Live* when Dana Carvey was in the cast.

As the box of paper clips falls, it accelerates (in this case, it speeds up), so we just use *F = ma*.

$$F = ma$$

Force only on box = (mass of box)(acceleration of box)

$$\frac{G(\text{mass of Earth})(\text{mass of box of paper clips})}{(\text{radius of Earth})^2} = (\text{mass of box})(\text{acceleration of box})$$

Now think back to the teeter-totter idea. The equals sign in *F = ma* means that the two sides balance. As long as we do the same thing to both sides of the equation (teeter-totter), we'll still have things in balance. The thing we're going to do is get rid of the mass of the box of paper clips from each side of the equation. We do that by dividing both sides by the mass of the box of paper clips (you might remember this as "canceling" from your math phobia days). If that seems like magic, well—it is. Okay, not really. Anyway, what you end up with is:

$$\frac{G(\text{mass of Earth})}{(\text{radius of Earth})^2} = (\text{acceleration of box})$$

The acceleration of the box of paper clips is given by:

$$(\text{acceleration of box}) = \frac{G(\text{mass of Earth})}{(\text{radius of Earth})^2}$$

This acceleration doesn't have *anything* to do with the mass of the box of paper clips! That means you could use the mass of anything you want and come up with the same answer, because the mass of the object always cancels out of the equation. So all things that are falling under the influence of gravity alone have the same acceleration, which turns out to be 9.8 meters per second per second (32 feet per second per second).

Now the tricky thing about all things having the same acceleration when they fall to Earth is that they rarely do that. To convince yourself, drop a plain sheet of paper and a crumpled sheet of paper side by side. The crumpled sheet hits the ground first, by a long shot. That's because pesky little things like air molecules are getting in the way, and our *F = ma* equation needs a force of air friction added to the left hand side. (Remember that the *F* in *F* = *ma* represents the total effect of *all* the forces acting on something.) We're back to that situation in Chapter 1 where everyday experience doesn't always seem to jibe with the scientific theory. So if you want to see how things work without air friction, head to the Moon where there isn't any air friction. You'll find that the crumpled and uncrumpled sheets of paper fall at the same rate. Or, instead of going there yourself, just ask Neil Armstrong. He performed a similar experiment when he was on the Moon, and I'm sure he wouldn't lie about the results.

4. **Are astronauts really weightless when they orbit the Earth? If not, why do they need those special toilets?**

What an absolutely stupid question. Anyone who's seen those space shuttle videos knows the astronauts are weightless. They float around in the cabin, their tubes of oh-so-tasty pureed food float around the cabin, and yes they do have to use special toilets or suffer dire consequences (they also use diapers—really). They're in the same situation as if they were standing on the Earth and some out-of-it technician turned off the force of gravity. So the astronauts *feel* weightless, but if you remember that weight is the *force of the Earth's gravity* on something, then you might realize that the astronauts still have weight. The force of gravity is weaker on astronauts because they're farther away from the center of the Earth (the *r* in Gm_1m_2/r^2 is larger), but—get ready for a big shock—no technician turned off the force of gravity. If you stick to the "force of Earth's gravity" definition, then you can't go anywhere in the universe and be weightless!

Okay, so why do astronauts *feel* weightless? It's because as they orbit the Earth or float toward the Moon, they're actually in free-fall. More about that in Chapter 7. For now, think of how you feel when you go over a crest on a roller coaster and your stomach seems to fly up; or when you reach for the top step of a staircase, realize it isn't there, and "fall" a bit; or imagine the feeling you'd have if the elevator you were in suddenly dropped out from beneath your feet. If you and everything around you are in free-fall, it's the same as there being no gravity at all. But technically you still have weight, and so do those astronauts.

5. **How does the Moon cause tides? Isn't it supposed to have something to do with gravity? How many high tides are there in 24 hours? (You can look this one up.) Does the Moon have to be full to cause high tides?**

Since you only had to use your brain in the previous "things to do" section, it's only fair that I make you *do* something in the reading section. Grab three rubber

Figure 4.5

bands, all the same size and thickness. Now find a block of wood and pound two nails into it so the rubber bands will stretch tightly between the nails (Figure 4.5).

Pull really hard to the side on one of the rubber bands. Then pull on a second one, but not as hard as the first. Finally, pull the third one to the side, but not as hard as the other two. (This last step might require a lovely and talented assistant.) You should end up with something like Figure 4.6.

Figure 4.6

By pulling on the rubber bands with different forces, you've separated them. I know, you can barely contain your excitement. But you've created a model for what causes the tides. In what follows we'll concentrate on the Moon and the Earth, but just so you know, the Sun also affects tides.

You know that everything exerts a gravitational force on everything else, so it shouldn't surprise you that the Moon pulls on the Earth and everything on it.

Figure 4.7

Figure 4.8

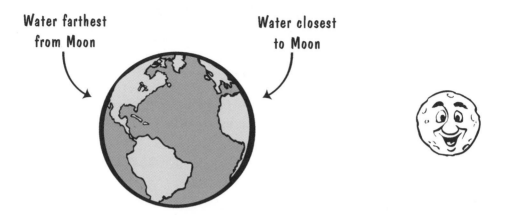

Water farthest from Moon

Water closest to Moon

Let's focus on the Earth, the ocean water closest to the Moon, and the ocean water farthest from the Moon. Gravitational forces get weaker the farther apart the objects. So the Moon exerts forces on things as follows:

Gravitational force of the Moon acting on the ocean water closest to the Moon→ strongest force

Gravitational force of the Moon acting on the Earth→ medium force

Gravitational force of the Moon acting on the ocean water farthest from the Moon→ weakest force

Let's see—three forces of different strengths—acting on three things that were together—reminds me of—hey, rubber bands! Because ocean water is free to move around the Earth's surface, the Earth and the ocean water on either side sort of get separated, producing a bulge of water on either side of the Earth.

Now, as the Earth spins, it exposes different parts of the oceans to the

Figure 4.9

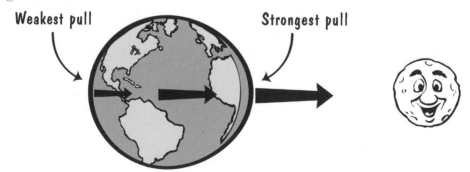

Weakest pull

Strongest pull

bulging effect. It's as if the bulges stay put while the Earth turns underneath. As your part of the Earth moves underneath a bulge, the water there is at high tide. How often does this happen? Well, there are two bulges of water (remember, the water is pulled away from the Earth on the side closest to the Moon, and the Earth is pulled away from the water on the side farthest from the Moon) and the Earth spins on its axis every 24 hours. Two bulges in 24 hours makes for a high tide every 12 hours (higher math). But you already knew that because you looked it up when I told you to.

Does the Moon have to be full for this to happen? No way. Just ask a fisherman. The Moon's gravitational pull doesn't turn off when there isn't a full Moon. The size of high and low tides does change depending on the phases of the Moon, but that's because the relative position of the Sun and the Moon, which affects the Moon's phases, also affects the size of tides. You'll have to trust me on that one, because I'm not going to explain the phases of the Moon right now. This chapter is too long already.

Figure 4.10

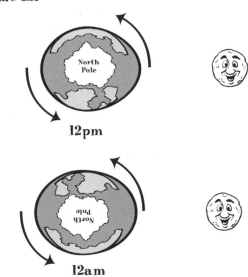

While the Earth rotates, the bulges stay where they are. As the Earth turns underneath the bulges, different parts of the Earth experience high and low tides.

Chapter Summary

- There is a gravitational force acting between any two objects that have mass.

- The gravitational force between two objects of mass m_1 and m_2 is given by $$F_{grav} = G\frac{m_1 m_2}{r^2}$$ where r is the distance between the centers of mass of the two objects.

- In the absence of air friction (not a common occurrence!), all objects near the Earth fall with the same acceleration, which is represented by g, and is equal to 9.8 meters per second per second.

- An object's weight is defined as the gravitational force exerted on the object by the Earth.

- An object's mass and its weight are not the same thing.

- Gravitational forces between common objects on the Earth's surface are so small as to be insignificant in our everyday lives.

- People who are in free-fall, accelerating only under the influence of gravity, experience a sensation of weightlessness. Technically, however, they do still have weight because the Earth is pulling on them.

- The tides are a result of the fact that the gravitational force an object exerts on other things gets weaker the farther you are from the object. Both the Sun and the Moon, along with the rotation of the Earth, are responsible for the Earth's tides.

Newton's Third

We have to visit our buddy Newton one more time in order to get a more complete under-standing of how and why things move the way they do. If you've been bugged about inanimate objects pushing on things, fear not. I'm finally going to explain how that happens. You'll also learn how rockets work and find out why most of the people in the world don't have a clue when they talk about action and reaction. Wooo hooo!

SCI LINKS
THE WORLD'S A CLICK AWAY

Topic: Isaac Newton

Go to: *www.scilinks.org*

Code: SFF10

Newton's Third

Things to do before you read the science stuff

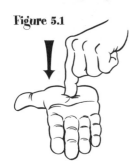

Figure 5.1

Hold your right hand out flat. Now take a finger from your left hand and press in the center of the palm in your right hand. What happens? If you say that you get an indentation in the palm of the hand being pressed, give yourself an A⁺ for this activity.

Try to make the palm of your hand indent *without* pushing on it with your finger or anything else. Telekinesis might help here.

Figure 5.2

Now use your palm to push against the corner of something that's unlikely to move, such as a table (Figure 5.3). Make sure the corner's not too sharp, okay? As you push, notice what happens to the palm of your hand. For those paying minimal attention, that would be the palm of the hand that's doing the pushing. Do you notice any similarity between what happens to your palm now and what happens when you push with the finger of your other hand? You should.

Finally, find something that will slide across the floor when you push it—a chair, a box, a reluctant two-year-old. Push this object with the palm of your hand and notice what happens to your palm *as you're pushing* (Figure 5.4). This takes a bit of concentration, so scout out your path before you start.

Figure 5.3

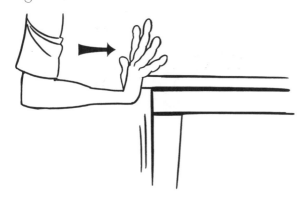

Figure 5.4

THIS
END
UP

I lied. That wasn't the final thing to do. Go find a bed and push on the mattress. Can you feel the mattress pushing back? Why is it doing that? While you're pondering, lie down and rest awhile. Science is hard work.

Figure 5.5

The science stuff

The palm of your hand won't indent without something pushing on it. Agree? If so, then answer this: When you use your palm to push on the corner of a table, is something pushing on your palm? Well, your palm indents, so my conclusion (also Newton's) is that the table is pushing on your palm. The idea of a table pushing on a person seems pretty silly to some people, but that's just because science concepts don't always hang out with common sense. I warned you about that in Chapter 1. What you have to do is quit thinking of forces as things that require some kind of conscious effort. That's easier to believe when you're pushing on a mattress. The springs in the mattress push back, and you can *feel* it.

Newton summed up this idea of objects pushing on each other and pushing back, and called it **Newton's third law**. Isaac said it like this: "To every action there is always opposed an equal reaction; or, the mutual actions of two bodies upon each other are always equal and directed to contrary parts." Probably a great pickup line in the seventeenth century. In everyday language, the third law says that when you push on an object, it pushes back with an equal force (equal in magnitude to

SCI LINKS.
THE WORLD'S A CLICK AWAY

Topic: Newton's third law

Go to: *www.scilinks.org*

Code: SFF11

Figure 5.6

Person pushes object

Object pushes back on person

your push) and that force is in the exact opposite direction from your push (Figure 5.6).

Most people have heard of Newton's third law as the **Principle of Action and Reaction**. That term is really misused, though. If I say "Boo!" and you jump, that's an action and a reaction. It also has nothing to do with Newton's third law. So the next time someone uses the words action and reaction to describe something that's just cause and effect, you can take comfort in knowing that they don't know of what they speak.

Now, even though you can see your hand indent when you push on a table, you might still be thinking that an inanimate object pushing on things is so much scientific horse pucky. If that's what you're thinking, go back and push on that mattress again (Figure 5.7). If it's a spring mattress, it's not hard to see why it's pushing back. As you push the springs away from their normal position, they have a natural tendency to "spring back" to that normal position. (That's why they call them springs!) In doing so, they push on your hand.

In a microscopic view, a table-top acts a whole lot like a spring

Figure 5.7

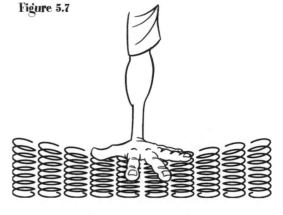

Figure 5.8

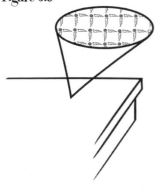

mattress. It's made up of a bunch of tiny things called atoms, and while those atoms aren't connected by springs, they interact through electric and magnetic forces so it's just as if they *were* connected by springs (Figure 5.8).

When you push on the tabletop, the electric and magnetic forces act just like springs and "push back" on your hand.

What? You don't believe in atoms because you've never seen one? No problem. Remember that all of science is *made up*. As long as the model (things made of atoms that interact through invisible forces) helps explain things, that's good enough. Heck, I don't really know why inanimate objects push back—they just do.

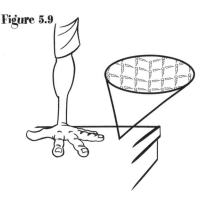

Figure 5.9

Electromagnetic forces act like springs and push back

More things to do before you read more science stuff

Find a place outside where you can walk and run, preferably alone because you're about to look foolish. First just walk normally and pay attention to what your body is doing. Pay special attention to how you go about starting from a standstill. Now run and do the same thing. What do you do to begin running from a standstill? Now answer a question. In going from a standstill to walking, or from a standstill to running, you're accelerating, right? (If you don't know why you're accelerating, head back to Chapter 2 for a review.) What *force* is causing that acceleration? In answering that question, remember that when *you* accelerate, that's caused by something *other than you* exerting a force on you.

Grab a rock or ball and hold it in your hand. Can you feel the rock exerting a force on your hand? (Say yes.) Can you see the effect of the rock exerting a force on your hand? (Notice the depression of your skin and say yes again.) What is the *net force* acting on the rock while it's resting in your hand? (Check out Chapter 3 for a definition of net force.) Now drop the rock onto your hand. When the rock hits, is it exerting a larger or smaller force than before? What about the force your hand exerts on the rock?

Figure 5.10

You push on the ground

The ground pushes back

More science stuff

Let's start with the walking and running. If you decided that the force causing you to go from a standstill to walking or running was caused by the *ground*, then you get first prize (Figure 5.10). Yeee ha! It's just

Newton's third law. You push on the ground, and the ground pushes back using the force of friction.

Of course, you're the one initiating all of this. It wouldn't do for the ground to just start pushing on you because it decided you needed to get moving. But without the ground pushing back, you wouldn't go anywhere. To see that I'm not lying, head out to your local ice rink or frozen pond and try walking, not skating, across the ice. It's really hard because there's very little friction between you and the ice (ice isn't great at grabbing on to passing objects—things slide by easily). While you're on the ice, strap on some skates and find a friend who also happens to have skates. Stand in the middle of the ice with your friend's back to you. Push on your friend's shoulders. Your friend moves away because you pushed him or her, but *you* begin moving backwards because your friend's shoulders obeyed Newton's third law and pushed back on you.

Figure 5.11

Push on your friend's
shoulders

Friend's shoulders
push back on you

On to the rock you held in your hand. Because you read and believed Chapter 4, you know that there's a force of gravity acting on the rock, and this force is the rock's weight. If gravity were the only force acting, then the rock would accelerate downward. Since it's sitting there in your hand, you know it's not accelerating. So there must be a force that just cancels out the force of gravity—the force your hand exerts (Figure 5.12). The *net force* acting on the rock is zero.

Figure 5.12
Force from your hand

Force of gravity
(rock's weight)

If your hand exerted a force larger than the rock's weight, the rock would accelerate upward. If your hand exerted a force less than the rock's weight, the rock would accelerate downward (through your hand—ouch).

Figure 5.13

Force from your hand

When you drop the rock into your hand, you can feel that the force the rock exerts on your hand is larger than when you just hold it. *Throw* the rock into your hand and you'll really notice a larger force! What net force is acting on the rock now? Well, gravity is still acting and it's the same as before. But since the rock is hitting your hand harder than before, the third law says that your hand is now exerting a larger force on the rock. (See Figure 5.13.)

Force of gravity (rock's weight)

This means that there's a net force acting upward on the rock. The rock is accelerating upward. Does that make sense? But of course. In going from dropping to being at rest in your hand, the rock is changing its motion. Any change in motion is an acceleration. The force that's causing the acceleration is the difference between the upward force of your hand (the larger force) and the downward force of gravity. Once the rock comes to rest, it's no longer accelerating. The upward force of your hand now equals the downward force of gravity (the rock's weight). And now that you know that, you can get a good night's sleep.

Chapter Summary

- When object A exerts a force on object B, object B exerts an equal and opposite force back on object A. This is known as Newton's third law.

- Newton's third law applies even when one or both of the objects are accelerating.

- Newton's third law is useful for explaining how airplanes, jet planes, and rockets do what they do.

*SCI*LINKS®
THE WORLD'S A CLICK AWAY

Topic: forces of flight

Go to: *www.scilinks.org*

Code: SFF12

Applications

1. Bet you're wondering how airplanes and rockets work. Well, wonder no more. First we'll deal with propeller driven planes. As the propeller turns, it's shaped so that it pushes air toward the back of the plane. As the propeller is pushing the air, the air is pushing back on the propeller, as you well know because Newton's third law says that's what happens. Because the propeller is attached to the plane, the plane gets pushed forward.

Figure 5.14

Propeller pushes air

Air pushes back on propeller

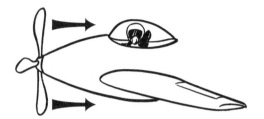

Can you fly an airplane to the Moon? No way. As you get farther and farther from the Earth's surface, there are fewer and fewer air molecules to push on. In outer space, there isn't *any* air to push on. If there's nothing for the propeller to push on, then there's nothing to push back and you can't accelerate.

Figure 5.15

Rocket pushes gases

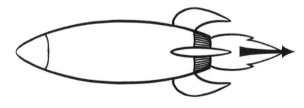

To head to the Moon you need a rocket, which uses Newton's third law in a different way. A rocket carries "propellants," which are basically things that explode when you mix them together and ignite them. The gases from the controlled explosions are forced out the rear of the rocket. In essence, the rocket *pushes* the gases out at a really high speed. The gases push back (third law) and the rocket accelerates forward (Figure 5.15). Because rockets push on these gases instead of pushing on air, they don't need no stinking air molecules and can cruise along just fine in outer space.

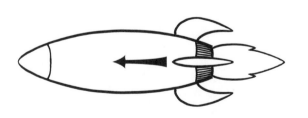

Gases push back on rocket

Now I know you're just dying to know how jet planes work. A jet engine has three main parts. (See Figure 5.16 and notice that we've moved on from a rocket to a jet plane.) The first is called a compressor. It contains a fan with

Figure 5.16

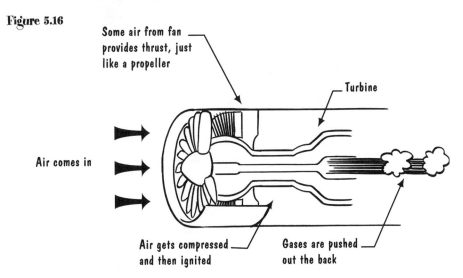

Some air from fan provides thrust, just like a propeller

Turbine

Air comes in

Air gets compressed and then ignited

Gases are pushed out the back

many blades that draw air in through the front of the engine. Next time you're close to a jet, look in the front of the engine (not too close!) and you'll see all the blades on that fan. In some kinds of jet engines, some of the air pushed back by the fan helps move the jet forward, just as with a propeller-driven plane. The rest of the air taken in is compressed and then mixed with fuel, which explodes. The gases from this continuous explosion are pushed out the back of the engine, providing a forward push in the same way a rocket engine does. On the way out, these gases cause a second set of fans, called a turbine, to turn. The turbine is connected to the first fan and keeps it turning.

Since a jet uses rocket-like propulsion, you might think it would work in outer space. *Au contraire,* my aviation neophyte. A jet engine relies on the air coming in the front to explode the fuel. Without this air, the fuel won't burn. No explosions means no gases pushed out the back and no outer space travel for this puppy.

2. When you watch a bunch of sprinters line up for the 100-meter dash, you'll notice that most of them use "blocks," which are thingies that are anchored in the ground that the sprinters push off from when they start the race. Do blocks really help? What would Newton say about it? Well, sure they help, otherwise sprinters wouldn't use them. At the start of the race, you want to get as big a push-off as possible (largest force pushing on you). The harder you push on something, the harder it pushes back on you (third law). If you

Figure 5.17

push *really* hard off the ground, chances are you'll slip—not a good start to a race. But if you push off something anchored into the ground so it won't slip, you can get as much of a push as your muscles will allow (Figure 5.17).

3. Okay, you're a smart person and, having read Chapter 4 only a short time ago, you're wondering whether Newton's third law applies to gravity. In other words, since the Earth pulls all sorts of things toward it, do all those things pull back on the Earth? Yep. Then why do things fall to the Earth instead of the Earth falling toward them? Newton's second law to the rescue (ta daa!).

    We'll write $F = ma$ for a falling object such as a turkey (turkeys can't fly, so it makes sense that they would fall).

(Force of Earth acting on turkey) = (mass of turkey)(acceleration of turkey)

Knowing that the turkey exerts an equal and opposite force on the Earth, we can write $F = ma$ for the Earth:

(Force of turkey acting on Earth) = (mass of Earth)(acceleration of Earth)

By Newton's third law, the forces on the left hand sides are the same force. The mass of the Earth is many, many, many, many times larger than the mass of the turkey, so the two teeter-totters (I introduced the teeter-totter idea back in Chapter 3) look like this.

Figure 5.18

So even though everything the Earth pulls to it also pulls on the Earth, the effect on the Earth is so tiny as to not be measurable. Also, the pulls of all these objects on all parts of the Earth tend to cancel each other out. That would have to do with the fact that the Earth is round, so that all objects on its surface are pretty much the same distance from the center of the Earth.

Figure 5.19

Objects all around the Earth pull it in different directions and the pulls tend to cancel each other out

Round and Round and Round in the Circle Game

If you recognize where the title of this chapter came from, then you're show-ing your age and your taste for FM music in the old days. For the record (and it *was* a record), it comes from a song written by Joni Mitchell. Anyway, this chapter is all about motion in circles. Although I've mentioned it a few times, you probably haven't noticed that just about everything we've talked about to this point has to do with motion in a straight line. When you move in a circular path, some strange things happen. Ooooooh—vewy scawy.

Things to do before you read the science stuff

Get about a meter of string and tie one end around something that's not so heavy that it would really hurt someone or something if it whacked them in the head. A roll of masking tape would be fine, a piece of lint isn't heavy enough, and the ugly brass paperweight you got for your birthday is way too heavy.

Figure 6.1

Figure 6.2

Go outside and start twirling the object over your head (Figure 6.2). While you're twirling, ask yourself this question: Is the object accelerating? Keep twirling while you look back at Chapter 2 to remind yourself what acceleration is. If it's accelerating, what force or forces are acting on it? Now predict what will happen when you let go of the string. Yeah, yeah, it's going to fly away—duh—but in what direction? At the point you release it, will it fly directly away from you, toward you, or in some other direction? (See Figure 6.3.)

Okay, try it. Let go of the string just as the object passes a certain point. Do this several times until you're sure of the direction the object goes when you release the string.

Lay the object and string out on the ground. Hold the free end of the string and try to pull the object toward you. More easy science. While still holding the end of the string, try to *push* the object away from you. Silly thing to try, yes? Now see if you can hold on to the end of the string and cause

Figure 6.3

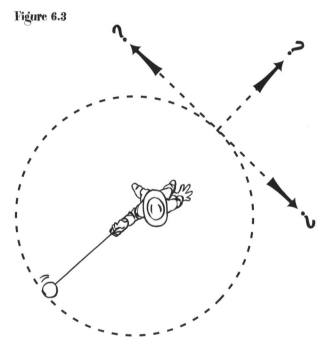

the object to move to your right or left. No fair moving the string over so you're *pulling*—you have to keep the string end where it is.

Figure 6.4

Try to use the string to
move the object this way

No fair moving over here

The science stuff

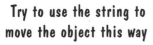

Is an object that's moving in a circle accelerating? That shouldn't be too tough a question because we answered it in Chapter 2. An acceleration is any change in speed and/or direction. Something going in a circle is definitely changing direction, so it's accelerating. Because we know the object is accelerating, we also know there has to be a *net force* acting on it (back to Chapter 3 if you forget what a net force is). Okay, so what's doing the pushing or pulling on your object attached to the string? The Earth pulls on it, but it should be obvious even to the most casual observer[1] that gravity isn't causing the object to move in a circle when you twirl it in a horizontal circle. So we'll forget about gravity.

What else can exert a force on the object? You? Not directly. What's touching the object? The string. Okay, you're pulling on the string and the string is pulling on the object, so in a way you're pulling on the object because you're causing the whole thing. But since the string is touching the object, we say the string is doing the pulling. In what direction is that force? Directly along the

---

[1]. I just had to include this phrase because authors of physics textbooks use it a lot. It's a condescending way of letting you know that you're something of an idiot if you don't get what the author is talking about.

Figure 6.5

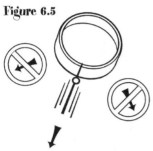

You can only use the string
to pull along the string

Figure 6.6

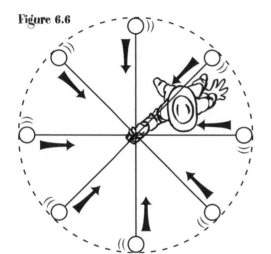

string. You know that because you convinced yourself earlier that you can't use a string to do anything but pull an object *directly toward you along the string* (Figure 6.5).

Anything else pushing or pulling on the object? If you ignore air friction (and gravity), the answer is no. So what we've got is something moving in a circle because it's being pulled toward the center of the circle.

That might seem weird but it's true. At this point you're probably wondering why the direction of the force is so important. If so, keep wondering until the next chapter. You also might be wondering why I haven't mentioned "centrifugal force." Seems just about everyone remembers from elementary school that there's a centrifugal force when something moves in a circle. You'll only have to wonder about that until we get to the next section in this chapter.

All right, in what direction did the object fly off when you let go of the string? If you do it lots of times and are precise about when you let go, you'll find the object takes off as shown in Figure 6.7.

The object keeps on moving in the direction it was going the instant you let go. For

Figure 6.7

Let go here and object
takes off in this direction

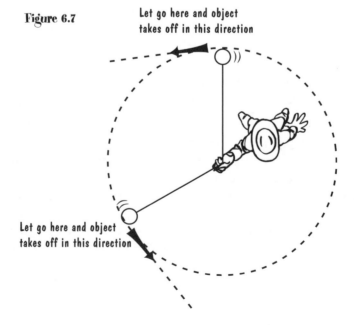

Let go here and object
takes off in this direction

all you wild and wacky Newton's first law buffs out there, this is a good example of it. The moment you let go of the string, the string no longer exerts a force on the object. Because the string was the only thing exerting a force in the first place (remember, we're ignoring gravity in this one), there is now *zero* force acting on the object. Zero force means zero change in motion. No change in motion from that instant on means the object just keeps on doing what it was doing, which is moving in a certain direction.

So here's how you can look at motion in a circle. At any instant, the object tends to keep doing what it's doing in a straight line, but you (or the string, or whatever) keep pulling on it and changing its direction. The result is a circle.

Figure 6.8

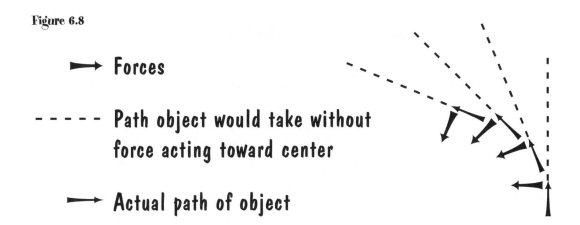

➡ Forces

- - - - - Path object would take without force acting toward center

➡ Actual path of object

More things to do before you read more science stuff

Grab a friend who knows how to drive and have him or her chauffeur you around the neighborhood. Take a ball with you. Place the ball on the dashboard or on the back seat and watch what it does as your friend starts and stops and goes around curves. (For those of you smart enough to live in the wide open spaces rather than a neighborhood, find an open field.) Nothing new here, because you already know about Newton's first law. When the car does something other than keep moving in a straight line (in other words, it accelerates), the ball tends to keep doing what it's doing and so rolls all over the place as far as the car is concerned.

Now close your eyes as your chauffeur takes some turns really fast. Pretend you don't know you're in a car and you think you're sitting on your couch at

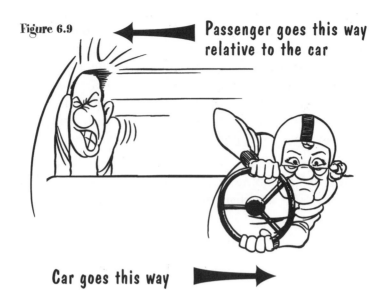

Figure 6.9

Passenger goes this way relative to the car

Car goes this way

home. Come up with an explanation for why you're getting thrown all over the place (Figure 6.9).

Go to the basement and dust off that phonograph turntable you don't use anymore. For you GenXers, a turntable is something that goes round and round at different speeds. Your parents used to place chemically processed petroleum products on a turntable, place a needle on top, and then do various things that they might or might not tell you about. If you don't live near your parents, the best way to scrounge a turntable is to wake up at 6:00 A.M. on a Saturday morning and cruise the garage sales. Be careful, though. Break-of-day garage salers will break your kneecaps before letting you have first pick of the items. If you can't find a turntable, you can use a lazy Susan (again, ask your parents what that might be) and turn it by hand for what follows.

Figure 6.10

Action figures at varying distances from the center of turntable
(Hey, you have your action figures, I have mine.)

Once you have a turntable, gather up three or four small action figures (one to two inches high). You can use anything about the same size and weight as substitutes for the action figures, but the aesthetics of what you're about to do just won't be the same. Place the action figures on the turntable at varying distances from the center—one right next to the center, one at the outside edge, and others in between.

Now turn the turntable on at its lowest speed. Really old turntables have speeds of 16 rpm, 33 rpm, 48 rpm, and maybe even 78 rpm. Newer ones probably just have 33 rpm and 45 rpm. Just start at the lowest number. Note which, if any, of the action figures falls off. Increase the speed of the turntable and see which ones fly off. Provide your own sounds of agony as the figures fly off. Do this several times. While noting which action figures fly off first, also check that they don't fly straight out away from the turntable, but along a path they're moving in at the instant they "lose their grip."

Figure 6.11

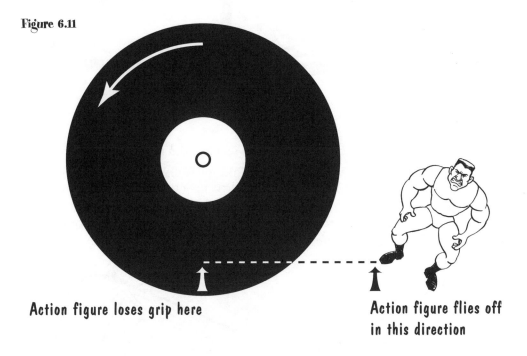

Action figure loses grip here

Action figure flies off
in this direction

Field trip time. Find a playground that has a merry-go-round. Like phono-graphs, merry-go-rounds are getting harder and harder to find. Your best bet is an elementary school that was built in the 60s and hasn't ever upgraded the playground equipment. Once you find a merry-go-round, you'll need an assistant and maybe some Dramamine. What you're going to do is take the place of the action figures that were on the turntable. Get on the merry-go-round and have your assistant twirl you at faster and faster speeds. How does the speed affect how hard you have to hold on to keep from falling off? Change your distance from the center and see how that affects things. Get off and throw up if you feel the urge. And while you're throwing up, may I reinforce the idea that actually experiencing the things I'm talking about (the activity itself, not the throwing up) will help immensely in your understanding of these concepts.

Figure 6.12

More science stuff

We're going to look at each of the activities you just did from two points of view. The first point of view is as an outside observer watching what's going on. You're standing on a road and a car comes by and turns left. A ball on the seat, or a person sitting in the car, moves to the right side of the car. Why? We already discussed that. The ball or person tends to keep on doing what it's doing (moving in a straight line) and the car accelerates out from underneath it (the car turns left).

Figure 6.13

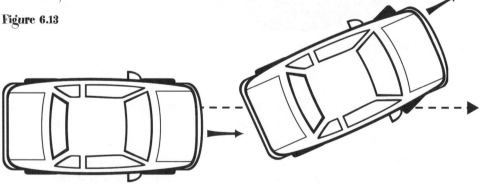

Car turns but ball tends to keep moving in a straight line

You don't need any strange, new explanation for what happens to the ball. Certainly you don't need to invent any forces that might have caused the ball to do what it did. In fact, the ball keeps moving in a straight line because there *aren't* any forces acting on it.

From inside the car, though, things look different. It seems as though the ball gets *pushed* from one side of the car to the other. If you were able to convince yourself you were sitting on your couch at home instead of sitting in a car turning left, about the only explanation you could give for your motion was that some unseen force was pushing you to the right. In fact, that's the explanation that scientists came up with.

Whenever you're in a situation where you're accelerating, such as when you're moving in a circular path, certain "fictitious" forces appear. When you're moving in a circle, there's a fictitious force called the **centrifugal force** that pushes you away from the center of the circular path. That's the force that pushed you to the right side of the car when the car turned left. As you look at the diagram in Figure 6.14, keep in mind that the arrow labeled centrifugal force only exists for someone who is riding in the car.

SCI*LINKS*.
THE WORLD'S A CLICK AWAY

Topic: centrifugal force

Go to: *www.scilinks.org*

Code: SFF13

Figure 6.14

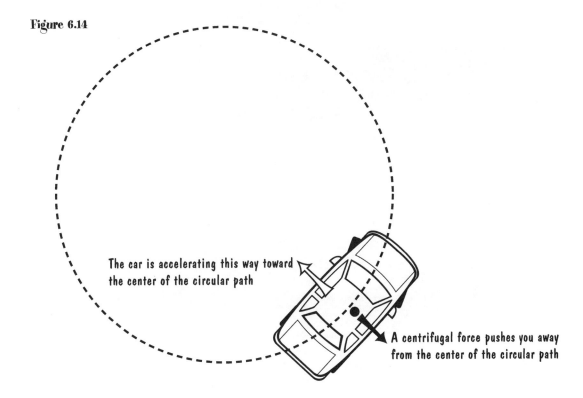

The car is accelerating this way toward the center of the circular path

A centrifugal force pushes you away from the center of the circular path

Because I'm using the term "fictitious force," some of you clever folks out there might be wondering whether or not these forces are *real*. You can't trace them to any object or person directly exerting them, but they are definitely real to the person riding in the car. Besides, if you haven't been paying attention up to this point, you might not realize that *all science concepts are made up*! In that sense, fictitious forces are as real as gravity or any other force. As long as they help explain how the world works, they're as real as any other science concept. It is important to realize, though, that fictitious forces only exist from the point of view of the person or object that is moving in a circular path. For someone watching from the outside, fictitious forces do *not* exist.

We can explain the turntable and the merry-go-round from two points of view, also. As an outside observer, you see the action figures accelerating (moving in a circle) along with the turntable as long as the friction between them and the turntable holds them there. It's just like an object on a string, with the friction taking the place of the force exerted by the string.

Figure 6.15

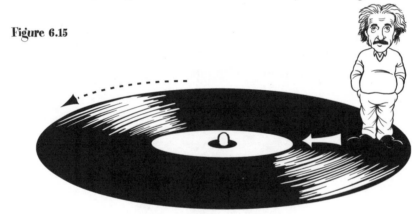

Friction between the action figure
and the turntable causes the action
figure to move in a circle

Figure 6.16

Once there is no longer a frictional force
the action figure moves off in a straight line

When the friction can no longer hold the figure in place, it takes off in a straight line in the direction it was moving when it lost contact. From the action figure's point of view, though, there's a "fictitious" centrifugal force pushing it out away from the center of the turntable. While the figure is still on the turntable, there's a balance between the frictional force and the centrifugal force.

As the turntable spins faster, the centrifugal force apparently gets larger, because at some point the centrifugal force is larger than the frictional force and the action figure slides off the turntable.

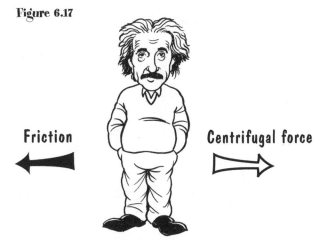

Figure 6.17

Friction   Centrifugal force

These forces are equal while the figure is on the turntable

When you're on a merry-go-round, you've got the same two points of view. From an outsider's point of view, you have to hang on to the merry-go-round in order to accelerate along with it. Actually, you holding on isn't what accelerates you. It's the force the merry-go-round exerts on you (third law—you push on the merry-go-round and it pushes back). The faster the merry-go-round goes round, the greater the acceleration. That requires a larger force, so you have to push harder on the merry-go-round so it pushes harder on you.

Figure 6.18

The merry-go-round pushes on you, keeping you moving in a circle

From your point of view on the merry-go-round, there's a balance between the centrifugal force pushing you out and the force of the merry-go-round pushing you in.

Figure 6.19

Centrifugal force   Push of the merry-go-round

As the merry-go-round speeds up, you have to hold on harder. That's because the centrifugal force gets larger as the merry-go-round spins faster.

You've got more evidence that the centrifugal force increases with speed. First, you have to realize that the farther from the center of the turntable or merry-go-round you are, the faster you move. In one revolution, each of the points 1, 2, and 3 end up back where they started. But it's obvious from the drawing (even to the most casual observer!) that point 3 travels a much greater distance than does point 1 in that one revolution.

Figure 6.20

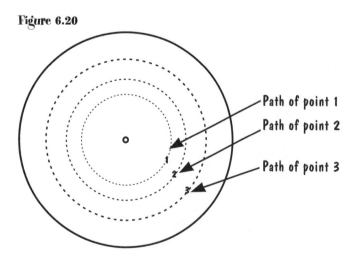

Path of point 1
Path of point 2
Path of point 3

When you speed up the turntable, the action figures closest to the edge fall off first. That means the centrifugal force is greatest at the edge, which is the part of the turntable that moves the fastest. Greater speed, greater centrifugal force. Your experience on the merry-go-round should give you the same result.

Even more things to do before you read
even more science stuff

Get a large sheet of paper (8½ x 11 is okay, but something larger would be better) and a felt-tip marker. Place the paper on top of the turntable you still have lying around. Push the paper over the center spindle and then tape it down. Use scissors to trim off any paper that's hanging over the edge of the turntable.

Figure 6.21

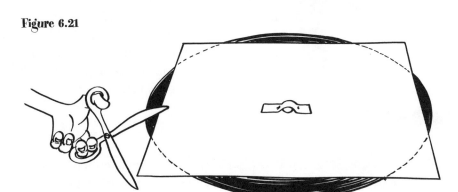

Start the turntable at its slowest speed. While it's turning, use the marker to draw a line (on the paper!) from the edge directly in to the center. Then draw a line from the center directly out to the edge.

Figure 6.22

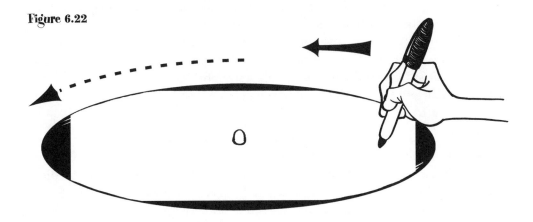

Figure 6.23

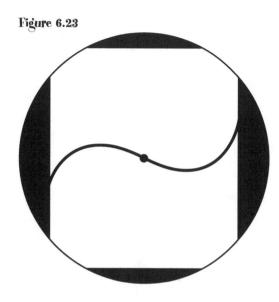

Stop the turntable and check out your "straight" lines. They should look something like Figure 6.23.

Head out to that merry-go-round again. While it's spinning, try walking from the edge into the center and then from the center out to the edge. Not so easy, eh? Try to figure out which way you're being pushed as you do this. If the motion doesn't bother you too much and you have a friend handy, sit on opposite sides of a merry-go-round and try to play catch with a ball. Nearly impossible.

Even more science stuff

SCILINKS
THE WORLD'S A CLICK AWAY

Topic: Coriolis force

Go to: *www.scilinks.org*

Code: SFF14

I'm guessing it was no big surprise that you ended up with curved lines when you tried to draw straight ones on a spinning turntable. After all, the turntable was moving underneath your pen as you drew. But suppose now that you're an ant trying to go directly from the edge to the center. You end up taking a curved path because you get pushed to the side the whole time! If you did as I asked and tried walking on the merry-go-round, then you experienced what the ant would have felt. What you felt was another fictitious force known as the **Coriolis force**. It's a very real force that arises when you try to move on something that's moving in a circle, and it pushes you to the left or right of your direction of motion, depending on which way the thing you're moving on is spinning. The result is that the Coriolis force tends to push you in a circular path when you're trying to walk a straight line. Just like the centrifugal force, the Coriolis force is only real for objects and/or people that are moving in a circular path.

Because the Earth is one of those things that spins all the time, there's a Coriolis force for everything that moves across the Earth's surface. We don't notice it in everyday situations, but it's responsible for wind patterns, the direction of spin of hurricanes, and the overall curved path of hurricanes. By the way, there's a common myth that bathtubs and toilets drain in one circular direction in the northern hemisphere and a different circular direction in the southern hemisphere. That would be true if the Coriolis force were the only thing that governed the direction the water swirls, but it's not. Same thing goes for tornadoes. Experiment for yourself to see that you can make the bathtub drain in either direction, depending on how the water is moving when you open the drain. Now *there's* a good use of your spare time.

Chapter Summary

● Objects that are moving in a circle, even at a constant speed, are accelerating.

● The force required to keep an object moving in a circle at constant speed is directed toward the center of the circle.

● You can describe the motion and acceleration of an object moving in a circle from two points of view. From a stationary point of view, watching the object move in a circle, there is no such thing as a centrifugal force or a Coriolis force. You simply determine what might be exerting a force on the object, causing it to accelerate in a circular path. From the point of view of the object, no acceleration is taking place. To explain what is happening from the object's point of view, we invent two "fictitious" forces—the centrifugal force and the Coriolis force.

Applications

1. The most important thing you can do with science is use it for magic tricks. This one isn't too magical, but it'll amaze anyone under the age of six. Get a bucket with a handle and fill it about a third full of water. Tell your audience you can turn the bucket upside down without any water spilling out. Take bets from the over-21 crowd. Then swing the bucket very fast in a vertical circle. As long as you swing it fast enough, the water will stay in the bucket. With a little practice, you can swing it rather slowly and still have the water stay in.

Figure 6.24

To see what's going on, look at it from the water's point of view, which means there will be a "fictitious" centrifugal force. At the top of the swing, gravity is pulling the water down. It's also possible that the bucket is pushing down on the water, keeping it moving in a circle. Because the water is moving in a circle, there's also a centrifugal force (at least from the water's point of view).

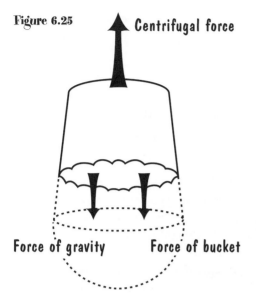

Figure 6.25

Centrifugal force

Force of gravity   Force of bucket

In order for the water to stay put and not accelerate downward and fall out of the bucket, the centrifugal force has to be large enough to cancel out the force of gravity and the force of the bucket. The faster you swing the bucket, the larger the centrifugal force. You can actually calculate the minimum speed (using $F = ma$ and a formula for centrifugal force that I didn't tell you about), but it's so much more fun to figure it out with trial and error. Wear a raincoat.

2. Ever wonder how they make those loop-the-loop roller coasters safe? Look at this from the point of view of someone sitting in the coaster. At the top of the loop, you've got the same situation as with the water in the bucket. The coaster needs to be going fast enough at the top of the loop that the centrifugal force is enough to keep you in your seat. In general, there could be a number of forces acting on someone in the coaster, including gravity pulling down, a centrifugal force pushing up ("toward the outside of the circle" is up when you're at the top of a loop), and any forces exerted on the person by the coaster itself.

Figure 6.26

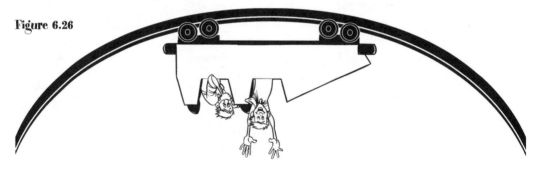

Centrifugal force

Force of gravity plus force of car

To find the minimum centrifugal force necessary to keep you in your seat, we set it equal to the force of gravity acting downward. That way, you'll have zero net force acting on you, and you won't accelerate downward (out of your seat!) even if the coaster doesn't exert any forces on you. By the way, if these two forces are equal, you are in free-fall, you feel weightless, and your stomach might be doing a flip or two.

At the bottom of a loop, you've got a different problem. There, the centrifugal force *adds* to the force of gravity, pushing you down. You (and the dog in Figure 6.27) feel as if some unseen force is pushing you down into the seat of the coaster. The human body can stand only so much force before blacking out (not good for amusement park PR), so the coaster has to be going slow enough or around a gradual enough curve that the centrifugal force doesn't get too large.

Figure 6.27

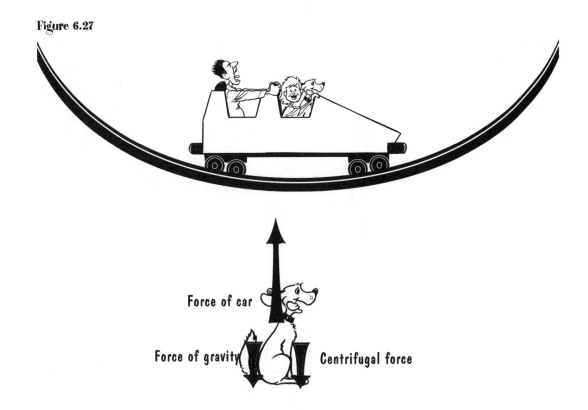

Force of car

Force of gravity   Centrifugal force

3. Keeping with the amusement park theme, what's going on with those rides where you spin in a circle, get pinned to the wall, and the floor drops out from beneath your feet? Below is a diagram of all the forces acting on you, from *your* point of view in the accelerating (spinning) frame of reference.

The centrifugal force doesn't directly affect whether or not you slide down the wall. In order to keep from sliding, the friction between you and the wall has to be as big as your weight (the force of gravity). The harder two things push together, the greater the friction between them, and this is where the centrifugal force comes in. The faster you turn in a circle, the greater the centrifugal force pushing you outward. This makes you push harder against the wall and vice versa (third law!). That increases the frictional force and keeps you where you are.

By the way, the friction force in Figure 6.28 is upward because gravity is tending to cause the person to slide downward. The frictional force opposes this motion.

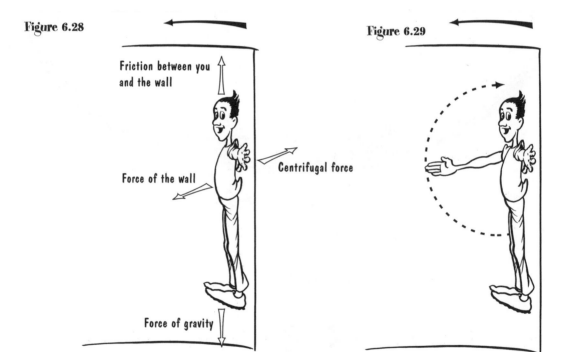

Figure 6.28

Friction between you and the wall

Force of the wall

Centrifugal force

Force of gravity

Figure 6.29

If you do hop on one of these rides, try moving your hand from your waist to your forehead, in sort of a tomahawk chopping motion (Figure 6.29). Can't do it. Know why? The Coriolis force (the one that pushes things to the side when they try and move in a straight line) comes into play. Ain't science neat?

Just for kicks, let's look at the person on the ride as viewed by someone who is stationary outside the ride. Now we don't need the fictitious centrifugal force to explain what's going on. We still have gravity and friction acting vertically, but the only force acting horizontally is the force of the wall pushing on the person. This force is directed toward the center of the circle, and is the force that keeps the person from flying off in a straight line, just as the action figures flew off the turntable. The faster the ride moves, the larger this force is. (It has to be larger because the acceleration is larger.)

Figure 6.30

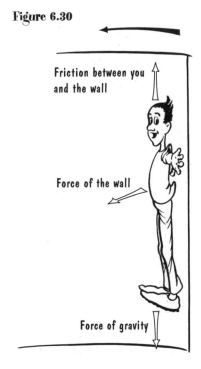

Friction between you and the wall

Force of the wall

Force of gravity

7
Chapter

To the Moon, Alice!

Y ou now know enough about force and motion to design a trip to the Moon. Yesireee, break out the champagne! Okay, so maybe you're not jumping up and down, but at least this chapter will give you an idea of how much you've learned. After all, if they can get a man on the Moon, you can understand science. We're not going to go into any detailed calculations, because that's not the purpose of this book. I just want you to be able to follow the general concepts.

This chapter doesn't have the same format as the others because there's only one activity for you to do. There are two sections. The second one is about getting to the Moon once you're in orbit. Before you do that, though, you have to think about—

Getting into orbit

First let's choose a spot from which to launch the rocket that's getting us to the Moon. Cape Canaveral is a handy place because they already have all the necessary equipment, museums, guided tours, and such. But suppose you wanted to build your own launch pad. Should you put yours in Florida, too? Keep in mind they have cockroaches the size of panda bears down there. Why not choose Maine so you can get great lobster and watch the fall foliage as you launch? Here's why you want to choose Florida. Take a look at a globe. The Earth spins on its axis. Which part of the Earth spins faster, the north pole or the equator?

Looking down from above the north pole, the Earth is just a 3-D version of a turntable. The farther out you are from the center, the farther you have to go in one revolution.

Figure 7.1

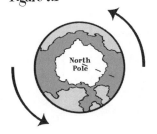

The closer you are to the equator, the farther you are from the center of rotation

So the equator is moving *much* faster than the poles. In fact, a point on the equator is cruising at about 1600 kilometers per hour (1000 miles per hour!). (Newton's first law explains why we don't notice that speed, by the way. We're in motion with the Earth and we tend to stay in motion with the Earth. It's just like riding along at a constant speed in a jet plane.) If you launch near the equator, you can get a big boost for heading into orbit (as long as you head in the right direction and not against the Earth's spin!) and you won't need as much fuel. Another way of looking at it is that the centrifugal force is greatest at the equator, so why not use that force to help you lift off? At any rate, you need a state close to the equator, so that's the choice to make. If you're more partial to Louisiana (crawfish and red beans and rice) or Texas (great barbecue), go ahead and put your launch site there.

Now that you're settled in the South, what about a launch pad? Do you really need one? After all, you've got to push off something, right? For a little insight, grab three friends and do the following:

Make two balloon rockets like the one in Chapter 3. No need for index cards this time, but you should make sure the two balloons are the same size and shape. Find a place where you can attach two strings, about a meter apart, to the ceiling. You also need to arrange things so one string hangs straight down over a table and the other off to the side of the table.

Figure 7.2

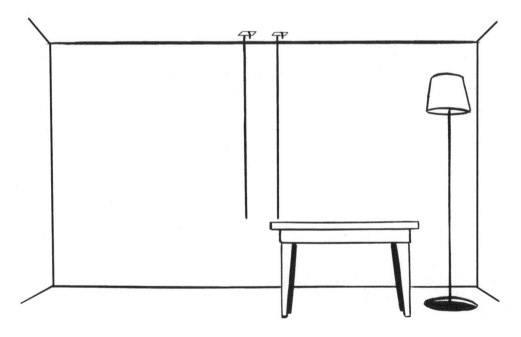

Put a balloon rocket on each string and aim them upward. Inflate the balloons the same amount. You'll need two people to hold the strings tight and two to hold onto the balloons so air doesn't escape. The nozzle of the balloon over the table should be just above the table. What you have is one rocket with a launch pad (the one over the table) and one without.

Let the balloons go *at the same time* and see what happens. Do this three or four times, and I'll bet the balloon rockets end in more-or-less a tie each time.

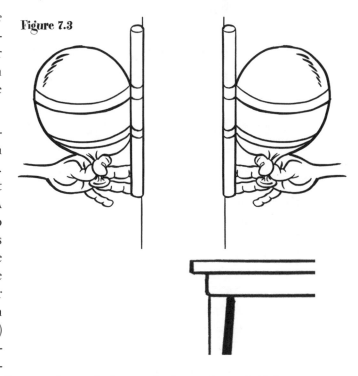

Figure 7.3

The fact that the two rockets tie must mean that a launch pad isn't really necessary. Maybe you figured that out before trying the activity. A balloon works according to Newton's third law. It pushes air out the opening, and the *air* pushes back, propelling the balloon forward. What the air does after leaving the balloon (such as hitting a table or not) has nothing to do with the motion of the balloon. The air already exerted a force on the balloon as it was being pushed out. Same deal for a rocket. The rocket pushes gases out the end and the gases push back. The launch pad has nothing to do with the rocket accelerating. Of course, a launch pad is a convenient place to put a rocket, and it prevents the rocket exhaust from scorching the ground, so you might as well go ahead and use one.

Time to worry about how much fuel to put in your rocket. If you run out of gas, it's a long walk to the nearest gas station. The first consideration is how far from the Earth you need to be to have a stable orbit. You have to get above the Earth's atmosphere so air molecules aren't continually slowing you down. Rather than go into a whole bunch of scientific stuff about how high the atmosphere is and how many air molecules you find at varying distances, we'll just choose a height. Okay, we won't just choose a height because as I'm writing this, the space shuttle is orbiting about 560 kilometers above the Earth's surface. Sounds like a good height to use. What we need to figure out is how fast the spaceship should be going so it can stay in orbit at that height. Hearken back to Chapter 6, where we viewed circular motion as motion in a straight line that gets changed by the force pulling toward the center.

Figure 7.4

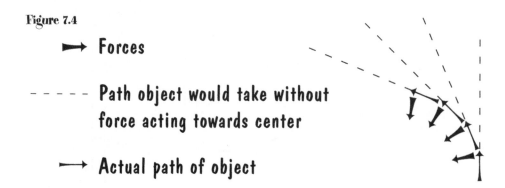

→ Forces

- - - - - Path object would take without force acting towards center

→ Actual path of object

As long as the speed is large enough, the pull toward the center will keep the object moving in a circle. Too small a speed for the given force and our space-ship spirals into the Earth because the force causes the ship to move in a tighter circle. Too large a speed for the given force, and the spaceship spirals outward. To figure out the necessary speed, we need to know how big the force is. What force are we talking about? Answer "gravity" and you win. How large is the force of gravity 560 kilometers above the Earth's surface? We just happen to have a formula for that (Chapter 5), but I'll save you any mathematical misery and tell you that 560 kilometers above the surface, gravity is about 5/6 of what it is at the surface. Not a big difference, but important if you want to stay in orbit.

A little aside here. This illustrates that you're far from being weightless (as long as we define weight as the force due to the Earth's gravity) when you orbit the Earth. It's just that you *feel* weightless because you and everything around you are in free fall.

Now that we know how big the gravitational force is up there, we can figure out how fast we need to go in order to stay in orbit. Correction: *I* can figure out how fast we need to go. I didn't clutter up your mind with a lot of formulas in the circular motion chapter, so you don't really know how to figure out the necessary speed. Trust me, though, it's just a simple application of Newton's second law. The result is something around 6400 kilometers per hour. Note to those of you with calculators and too much time: I just estimated that last number, so it's not exact. As long as the folks at NASA aren't as lax as I am, we'll be fine.

Okay, assuming you have several gazillion dollars to spend in order to make this trip a reality, you'll need to figure out how much fuel to take. I won't calculate exact amounts for you because frankly, I don't have all the necessary specifications for fuel consumption on a big ol' nasty rocket that will take you to the Moon. What I will do is lay out the important concepts so you can see that if we had all the necessary details, we could figure things out exactly.

You're starting out going about 1600 kilometers per hour at the Earth's surface, and you need to hit a speed of about 6400 kilometers per hour in orbit. Let's see—change in speed and direction—must be an acceleration. Using those formulas I introduced at the end of Chapter 2, you can take the initial speed, the final speed, and the time of acceleration to figure out just what magnitude of acceleration is necessary. Once you know the necessary acceleration, you can use $F = ma$ to figure out how big a force you need. Once you know the force, you can use all you know about how fast the rocket pushes gases out the back end to figure out how much fuel you need.

As you might expect, I'm oversimplifying. For one thing, the acceleration isn't in a straight line and it isn't constant. For another, the mass of the rocket keeps changing as you send gases out the back end (those gases used to be part of the rocket, and they have mass). And even small changes during flight can drastically affect how much force you get from the escaping gases. Fortunately for the folks at NASA, there's a form of mathematics called calculus that takes care of all the changes that are involved. Fortunately for you, I won't be doing any calculus in this book. What I hope is clear is that the basic concepts needed for getting into orbit aren't all that difficult to grasp.

One important thing to consider is that once you're in orbit, you shouldn't use much, if any, fuel. That's because, in order to move in a circle, you need a force acting toward the center of that circle. The Earth is really nice in that it provides that force—gravity. This is why I told you in the last chapter (I really did—go back and check.) that the *direction* of the force required to move in a circle was important. Of course, you're going to need more fuel if you're going to go—

On to the Moon

All right, you're in orbit and you have space sickness and you're already tired of freeze-dried food, so let's get to the Moon already. At what point should you hit the thrusters and break out of your orbit? Maybe you should take advantage of Newton's first law and do it at the point you're heading toward the Moon at 6400 kilometers per hour. Check out Figure 7.5.

Figure 7.5

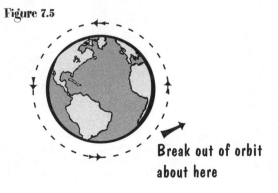

Break out of orbit about here

Didn't take a rocket scientist to figure that one out, did it?

Suppose you don't like that feeling of weightlessness and you want to create artificial gravity on the way. To get a clue as to how to do that, rent a copy of *2001: A Space Odyssey* and see how they do it. It's a great movie and only lasts about 17 hours, depending on how long you take for intermission. Check out the scene of Dave going for his morning jog.

Back so soon? Hopefully you figured out that HAL created artificial gravity by spinning the spaceship. When you're on a spinning ship, you feel a centrifugal force pushing you outward. If you adjust the speed at which you spin and take into consideration how big a circle you're going in, you can create a centrifugal force that exactly matches your weight on the Earth. So you live in a spinning doughnut with your feet always to the outside of the doughnut and your head toward the center.

Not so strange, actually, when you consider that on Earth we're doing pretty much the same thing with the directions reversed.

Figure 7.6

Okay, you're on your way to the Moon and you're not exactly sure when to cut off your engines. Is there some point at which you can just coast? Well, there's certainly a point at which the gravitational pull of the Moon equals that of the Earth. Get beyond that point and the Moon will pull you on in. Turns out that point is pretty easy to figure out. You just equate the Moon's pull to the Earth's pull. In case you don't remember the general formula for gravitational force that I introduced back in Chapter 4, it looks like this:

Figure 7.7

$$F_{grav} = G\frac{m_1 m_2}{r^2}$$

To set the Moon's pull on the spaceship equal to the Earth's pull on the spaceship, we just have to use this formula for the Moon-spaceship force and set it equal to this formula for the Earth-spaceship force. Because I don't want to end the book without one more set of equations, here goes:

gravitational pull of Earth = gravitational pull of Moon

$$\frac{G(\text{mass of Earth})(\text{mass of spaceship})}{\left(\text{distance from Earth}\right)^2} = \frac{G(\text{mass of moon})(\text{mass of spaceship})}{\left(\text{distance from Moon}\right)^2}$$

Solve it and get:

$$\frac{\text{distance from Moon}}{\text{distance from Earth}} = \sqrt{\frac{\text{mass of moon}}{\text{mass of Earth}}}$$

I skipped a whole bunch of algebra steps in getting to the second line, so don't get upset if you don't follow what I did. And in case it's been a really long time since you did any math, the symbol $\sqrt{}$ means "take the square root of."

Notice that, after you put in numbers for the mass of the Moon and the mass of the Earth, the result only tells you what you get when you divide the distance from the Moon by the distance of the Earth. It doesn't give you the exact distances. For an exact answer, you need to include the fact that it's about 384,000 kilometers from the Earth to the Moon. Knowing this number, you can figure out that the Moon's pull takes over when you're about 346,000 kilometers from the Earth.

A few extra points. First, notice that our earlier calculated speed of 6400 kilometers per hour is about right for an average for the trip.

$$\text{distance} = (\text{average speed}) (\text{time of travel})$$

$$\text{time of travel} = \frac{\text{distance}}{\text{average speed}}$$

$$= \frac{384\ 000\ \text{kilometers}}{6400\ \text{kilometers/hour}}$$

$$= 60\ \text{hours}$$

It took the Apollo astronauts around three days to get to the Moon, so 60 hours isn't far off.

Second, the point at which the gravitational pulls equal one another is on a line between the Earth and the Moon. In order to hit that point (and save on fuel, by the way), your spaceship should trace a figure-eight path rather than an elliptical one. The Apollo folks figured that one out, too.

Figure 7.8

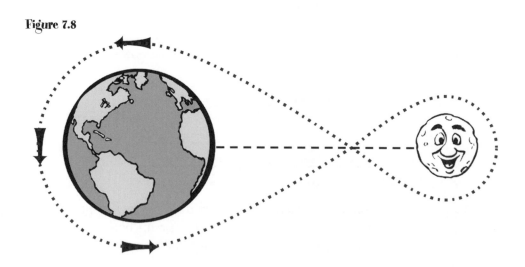

Finally, it turns out that you don't have to fire your engines all the way up to the point at which the Moon's pull takes over. You can cut them off much sooner and still reach that point. It's the same as throwing a rock into the air—you don't have to push it all the way in order for it to reach a certain height. A strong enough initial velocity is enough. We're not going to calculate the exact point at which you should shut off the engines because it involves that nasty old calculus stuff (which, like science, isn't as difficult to understand as many of the academics would have you believe).

Topic: rockets

Go to: *www.scilinks.org*

Code: SFF15

Well, I got you to the Moon—call if you can't get back okay. Here's hoping the journey was fun and that you learned a bit of science along the way. As you head off into the moonset, remember that science isn't that difficult to understand. I think I can, I think I can, I think I can—

Final comments on Newton's laws and reality

Newton's laws, and all of the other motion principles presented in this book, do a pretty darned good job of helping us understand, explain, and predict the world around us. We can use them to get to the Moon, design roller coasters, and develop safety regulations for cars. There are situations, though, where all the motion stuff in this book breaks down and is just plain wrong. One situation is when objects move at speeds close to the speed of light. Not your everyday occurrence, but quite common when you start accelerating small particles like electrons and protons and smash them into each other. You might not do that in your backyard, but physicists do it all the time, using particle accelerators. They have discovered a whole lot about the behavior of these tiny particles with such collisions, and have used the results to form new theories about how all things in the universe behave.

A second situation in which the contents of this book get thrown out the window is the study of all things astronomical. When dealing with gravity on a large scale, meaning very large masses such as planets and stars, some of the underlying assumptions of our basic ideas about motion don't accurately describe things. The theory of gravity presented in Chapter 4 doesn't quite work, and we have to start talking about curved space-time, expanding and contracting universes, and all sorts of cool things. As a side note, a person named Albert Einstein was primarily responsible for both of these corrections to the laws of motion. Smart guy.

This being the end of this particular book, we won't be getting into the complications Al thought up. Be on the lookout, though, for books in the *Stop Faking It!* series that address those topics.

Glossary

acceleration. Any change in motion of an object. A change in direction and/or speed. Accelerations are caused by net forces. Acceleration is a vector.

accelerator. That funny pedal on the right that people talking on cell phones have trouble finding when the light turns green.

Aristotle. A Greek philosopher who, among other things, had a view of motion more similar to Einstein's than to Newton's. Aristotle lived from 384 B.C. to 322 B.C. and was a major influence on the philosophy of his time. Rumor is he was a pretty good teacher.

average speed. The total distance traveled divided by the time of travel.

centrifugal force. A "fictitious force" that arises whenever an object is moving in a circular path, as seen from the object's point of view. The centrifugal force is directed away from the center of the circular path.

Coriolis force. A "fictitious force" that arises whenever you attempt to move on an object that's already moving in a circular path. The Coriolis force acts at right angles to your motion and makes it tough to walk a straight line if you're an ant on a turntable.

Einstein. A German-born scientist who, among other things, totally revolutionized Newton's theory of gravity. Don't look for Einstein's theory of gravitation in this book. He is popularly known for his *special theory of relativity,* which details the inadequacy of Newton's laws of motion at speeds approaching the speed of light. Einstein lived from 1879 to 1955.

force. A push, pull, whack, or bump; something you want to have with you when you're fighting intergalactic bad guys. Force is a vector.

friction. The force between two objects when they rub together.

g. A symbol for the acceleration due to gravity of everything that's freely falling near the surface of the Earth; $g = 32$ ft/sec^2 or 9.8 meters/sec^2.

G. The universal gravitational constant, equal to 6.67×10^{-11} m^3/kg-sec^2. This number goes in the formula for the gravitational force between two objects.

Galileo. A precursor to Newton who, among other things, experimented with the motion of objects. Galileo lived from 1564 to 1642, and is credited with the invention of the telescope. He used that invention to discover four of the moons of Jupiter.

Glossary

inertia. The tendency of an object to keep doing whatever it's doing. The inertia of an object is measured by its mass.

instantaneous speed. How fast something is going at any point in time.

magnitude. A number, usually applied to a vector, that tells how much of something there is.

mass. A measure of an object's inertia and the m that goes into $F = ma$.

net force. The sum total of all the forces acting on an object. Direction is important, so you don't just add the numbers together. Two or more forces can cancel each other with appropriate directions.

Newton. A guy who, among other things, stole some of Galileo's ideas and named them after himself. Actually, that's not fair to Isaac. He not only outlined the laws of motion, but did a considerable amount of pioneering work with light and optics. Newton lived from 1642 to 1727.

Newton's first law. The first of Newton's laws. A principle that states that objects tend to keep on doing whatever they're doing unless a net force acts on them.

Newton's second law. The second of Newton's laws. The relationship between an object's mass, its acceleration, and the net force applied to the object. You write it as $F = ma$.

Newton's third law. The third of Newton's laws. When one object exerts a force on a second object, the second object exerts an equal and opposite force back on the first. Also known as action/reaction. Also misunderstood by lots of people.

speed. How fast something is going, measured by the distance traveled divided by the time to travel that distance.

tides. The rise and fall of the oceans on a 12–hour schedule, caused by the gravitational pull of the Sun and Moon.

vector. A quantity that has both magnitude and direction. Force, velocity, and acceleration are all vectors.

velocity. The speed *and* direction something is moving. Velocity is a vector.

weight. The force the Earth's gravity exerts on something.

National Science Teachers Association